Beyond the Standard Model

Modern physics problems and solutions

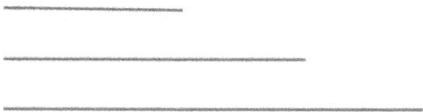

Includes empirical evidence and new approaches to the S. M.

Hossein Javadi

Title: **Beyond the Standard Model**

Subtitle: **Modern physics problems and solutions**

Author: **Hossein Javadi**

Publisher: **Supreme Century**, USA

ISBN: **978-1939123626**

Prepare for Publishing by
ASANASHR.com

This book is presented to:

My teachers and professors humbly taught the principles of humanity and morality more than school lessons. Thus, respecting them, I present this book to the teachers all over the world.

Note about the book cover image: In CPH theory particles are producing fields. The book cover image shows the birthplace of four fundamental interactions. Description in chapter 4 of the book.

Contents

Preface

Most of the materials in this book are from articles and excerpts from my point of view in the discussions that have taken place in scientific networks over the past few years, including: www.researchgate.net , www.linkedin.com and www.quora.com , has been raised. But there are a lot of explanations in this book that have never been published before. Completely, this book presents a brief description of the creative particles of Higgs theory or CPH Theory introduced by the author of this book in 1987. This theory is based on the generalization of light speed from energy to matter and vice versa.

In other words, CPH theory describes the physical properties of the least amount of energy, according to which the equation $\nabla V = 0$ was given in which V is speed. In the year 1992, I published a paper titled "Unify of Mass-Energy" in the journal of the secretariat of the faculty of Azad University, South Tehran, pages 15 to 17, and the book of scientific theories - rejection or generalization? Which was published by Eta Publishers in 1992. The last chapter of the book was devoted to explaining this theory as the theory of invariance. The invariance theory was meant to describe the same equation $\nabla V = 0$.

In 2006, I published the book Physics "from the beginning up today" by Eta Publishing, whose last chapter was devoted to the theory of CPH, in which I described the above equation extensively and using this equation, how to produce particles and fields I described the electromagnetic field, and for the first time, the mechanism of field production by particles was described.

At that time, the predictions of CPH theory were inconsistent with the principles of modern physics. Using the equation $\nabla V = 0$, which was why the reactions were very negative and disappointing. But I was not disappointed instead I published a large number of articles in scientific journals and I was looking forward to the future.

Until in the last two years, the results of physical experiments in internationally validated laboratories, and various papers on photon formation and its properties, quantum vacuum fluctuations, speeds above the speed of light, and ... by accredited physicists from different universities of the world It was published that, many years ago, they had raised and explained in my books and articles in more detailed. A list of some of my articles and the comparison of the predictions of CPH theory with the new experimental results and the new theories are given below in this preface.

The simultaneous publication of this book in Persian and English is a positive response to the expectations of those who, both from Iran and from different countries of the world, are demanding a coherent collection of CPH theory combined with empirical evidence and various theories that recent years have been presented to complete modern physics.

In the end, I would like to express my sincere gratitude to the distinguished professors, in alphabetical order: Dr. Parviz Tajdari, Dr. Aresu Jahanshir, Dr. Mostafa Shahzamanian, Dr. Farshid Forouzbakhs, Dr. Mahmood Quran Newis, for their help me to develop the CPH Theory.

Also I I would like to express my sincere gratitude to Dr. Hamed Daei Ksamei, Ms. Matin Vaez Tehrani and Mr. Amin Makfi for their assistans during the research and written books and articles.

Hossein Javadi

Tehran, Shahrivar 1396 / Augut 2017

Javadi_hossein@hotmail.com

http://cph-theory.persiangig.com

The address of the author's discussions in scientific networks:

https://www.researchgate.net/profile/Hossein_Javadi

https://www.linkedin.com/in/hossein-javadi-272a2b48/

https://www.quora.com/profile/Hossein-Javadi-1

Author's writings related to the subject of this book

1 - Unify of Mass-Energy, in Persian, 1992

2 – The scientific theories - rejection or generalization? In Persian 1992

3 – Physics: from the beginning up today, in Persian, 1992

4 - A New Mechanism of Higgs Bosons in Producing Charge Particles, General Science Journal, 2006,

http://gsjournal.net/Science-Journals/Research%20Papers/View/936

5 - Time Function and Absolute Black Hole, General Science Journal, 2006,

http://gsjournal.net/Science-Journals/Research%20Papers/View/942

6 - Zero Point Energy and the Dirac Equation , General Science Journal, 2007,

http://gsjournal.net/Science-Journals/Research%20Papers/View/950

7 - A New Definition of Graviton, General Science Journal, 2008,

http://gsjournal.net/Science-Journals/Research%20Papers/View/951

8 - New Discoveries and the Necessity of Reconsidering the Perspectives on Newton's Second Law, Journal of Nuclear and Particle Physics, p-ISSN: 2167-6895   e-ISSN: 2167-6909, 2012; 2(3): 31-35

doi:10.5923/j.jnpp.20120203.02

http://article.sapub.org/10.5923.j.jnpp.20120203.02.html#Ref

9- Sub quantum space and interactions properties from photon structure to fermions and bosons, Scientific Journal of Pure and Applied Sciences, Vol 2, No 5 (2013)

http://sjournals.com/index.php/SJPAS/article/view/678

10 - Definition of singularity due to Newton's second law counteracting gravity, Scientific Journal of Pure and Applied Sciences, Vol 2, No 3 (2013)

http://www.sjournals.com/index.php/SJPAS/article/view/602/pdf

11 - Interactions Between Real and Virtual Spacetimes, Fundamental Journals publication, .2014,

DOI:10.14331/ijfps.2014.330075

http://fundamentaljournals.org/ijfps/downloads/75_IJFPS_Dec_2014_114_121.pdf

Recently, empirical evidence and new essays have been published, with some of the predictions of the CPH theory are compatible. It is especially important for photon structure. Some of these articles have been reviewed in the following links and are consistent with the theory CPH. Comparing the date of publication of my works and these articles is noteworthy. Including:

1 – New reasonable evidence of CPH Theory's predictions

https://www.researchgate.net/publication/311962908_New_reasonable_evidenc e_of_CPH_Theory%27s_predictions

2 - Review and analyzing the evidence of the existence of quantum fluctuations

https://www.researchgate.net/publication/312586105_Review_and_analyzing_t he_evidence_of_the_existence_of_quantum_fluctuations

3 - If there are no particles and there are only fields, what are fields made up of?

https://www.researchgate.net/publication/312372328_If_there_are_no_particles _and_there_are_only_fields_what_are_fields_made_up_of

4 - Review the time-resolved scattering of a single photon by a single atom

https://www.researchgate.net/publication/311511253_Review_the_time- resolved_scattering_of_a_single_photon_by_a_single_atom

5 - A new mechanism of Higgs bosons in producing charge particles

https://www.researchgate.net/publication/237412049_A_new_mechanism_of_ Higgs_bosons_in_producing_charge_particles

Introduction

At the beginning of the 20th century, Newton's second law was corrected considering the limit speed c and the relativistic mass. The relativistic Newton second law which shows the mass variations (i.e., the infinite speed in classical mechanics is replaced by the infinite mass). In addition, infinite density appears in gravitational singularity. According to general relativity, the initial state of the universe, at the beginning of the Big Bang, was a singularity. Both general relativity and quantum mechanics break down in describing the Big Bang. At this time, the Big Bang, all the matter in the universe, would have been on top of itself. The density would have been infinite.

The approach of relativity toward the physical phenomena is hyper structural and explains the observations of the observer while there is little consideration to the intrinsic entity of the phenomena. In this book from a new approach to field, energy and matter, the relativistic Newton's second law is reviewed. According to this approach a new definition of singularity is predicated that shows in the singularity state: *volume will not be zero, density will be limited and even physical laws are able to* describe before the Big Bang and the reason of inflationary universe.

In quantum electrodynamics (QED) a charged particle emits exchange force particles (discrete amounts of energy) continuously. This process has no effect on the properties of a charged particle such as its mass and charge. How it is explainable? If a charged particle as a generator has an output known as a virtual photon, then what will be its input?

In relation to charged particles, a more fundamental question is raised: **How electrons absorb and emit light?**

What we know from quantum mechanics is: "when an electron absorbs energy, it jumps to a higher orbital. An electron in an excited state can release energy and 'fall' to a lower state."

In quantum mechanics, the concept of a point-like particle is complicated by the Heisenberg uncertainty principle, because even an elementary particle, with no internal structure, occupies a nonzero volume. According to the quantum mechanics that photon and electron are unstructured particle, we cannot answer the unanswered questions. But new articles and experiments show we need to review the concept of point-like, because electron and photon have structure.

Tehran 2017

Hossein Javadi

1

Photon - Graviton Interaction

1.1 The biggest problems in modern physics

The main unsolved problem in quantum field theory is how to unify the gravity and the quantum electrodynamics. Other problem is combining the quantum physics and general relativity. This problem sits on the fence between cosmology and particle physics: How can we merge quantum theory and general relativity to create a quantum theory of gravity? [1]

The root of the quantum gravity problem is that physicists want to solve the quantum gravity problem regardless the relativistic Newton's second law. While these two laws (universal gravitational law and relativistic Newton's second law) are closely related to each other. In addition, not only modern physics is involved such problems, but quantum field theory (which quantum gravity is a part of it) is involved with other problems. For examples; Do virtual particles come out of nothing? Do virtual photons possess mass or charge? How does a charged particle produce its own electrical field? In quantum field theory, the graviton is massless with a

---

[1] - John Baez, "Open Questions in Physics", 2012,
http://math.ucr.edu/home/baez/physics/General/open_questions.html

spin of 2 that mediates the force of gravitation. This is because the source of gravitation is the stress-energy tensor, a second-rank tensor[2].

But evidence show, gravitational potential energy is quantized. In fact, old definition of graviton is not able to solve quantum vacuum problem. According to Heisenberg's uncertainty principle, a vacuum isn't empty, and filled with particle-antiparticle pairs that appear and disappear randomly. So, we need a new definition of graviton that should be based on developing old theories and experimental evidences and resolve the renormalization problem.

1.1.1   Gravitational field

Given Newton's law of universal gravitation, every massive particle, even electron, creates its own gravitational field. Not only massive particles, even photons carry their own gravitational fields that are inherent in their mass-energies equivalence. The gravitational field of a massless point particle is first calculated using the linearized field equations[3]. In classical mechanic, the gravitational field g due to a point mass M is a vector field consisting at every point (with distance r of point mass M) of a vector pointing directly towards the particle that is given by;

$$g = \frac{GM}{r^2} \qquad (1\text{-}1)$$

G is th gravitational constant. For the gravitational field g around a light point mass m_{light}, we can write:

[2] - Don Lincoln, Gravitons, Fermilab Today, 2012

http://www.fnal.gov/pub/today/archive/archive_2012/today12-10-19_NutshellReadMore.html

[3] - P. C. AichelburgR. U. Sexl, On the gravitational field of a massless particle, Springer Link, 1971, https://link.springer.com/article/10.1007/BF00758149

$$g_{m_{light}} = \frac{Gm_{light}}{r^2} \quad (1\text{-}2)$$

A star is made up of atoms, each atom contains a few sub atomic particles, and each element has its own gravitational field. So, the gravitational field of a star is formed of combination the gravitational fields of its sub atomic particles. When a star explodes, every part of it such as sub atomic particles carries its own gravitational field. It shows sub atomics particles absorb each other, even in star. In the other word, gravitational field is quantized.

With regard to the exchange particles concept in the quantum field theory and the existence of graviton, when a particle / object is falling in the gravitational field, it goes from a low layer to a higher layer density of gravitons. Thus, we should investigate the impact of changing the density of gravitons on the exchange gravitons between the particles that in continue will be done.

1.1.2 Renormalization

The term renormalization comes from quantum field theory and is usually thought of as being the prerogative of that discipline[4]. Renormalization is a collection of techniques in quantum field theory that are used to treat infinities arising in calculated quantities that was first developed in quantum electrodynamics (QED) to make sense of infinite integrals in perturbation theory. The integrals for a particle of spin J in D dimensions is given[5] by:

$$I_{loop} \sim \int p^{4j} d^D p \quad (1\text{-}3)$$

For; $4J + D - 8 < 0$, the integral behaves fine for infinite momentum.

---

[4] - David McComb, RENORMALIZATION METHODS, CLARENDON PRESS, OXFORD, 2004

http://home.basu.ac.ir/~psu/Books/%5BW._D._McComb%5D_Renormalization_methods_a_guide.pdf

[5] - Why did strings enter the story, Superstring theory, http://superstringtheory.com/basics/basic3a.html

For photon; $D = 4$, $J = 1$, integrals are renormalizable.

For graviton $D = 4$, $J = 2 \rightarrow 4J + D - 8 = 4$, integrals are not renormalizable.

String theory has solved this problem with other approach on the problem[6]. In CPH theory, after reconsidering and analyzing the behavior of photon in the gravitational field, a new definition of graviton based on carrying the gravity force is given. First of all we need to focus on concept of photon and its properties. Because gravitational potential energy is a kind of energy that is convertible to electromagnetic energy and vice versa. So, true understanding the structure of photon is needed to describe the properties of graviton well.

1.2 What is a photon really?

Maxwell formulated dynamical theory of the electromagnetic field in 1864 as follow;

$$E_x(z,t) = E_{max}cos2\pi\left(vt - {}^z\!/_\lambda\right) \qquad (1\text{-}4)$$

$$B_x(z,t) = B_{max}cos2\pi\left(vt - {}^z\!/_\lambda\right), c = v\lambda \qquad (1\text{-}5)$$

Light is an electromagnetic wave. The speed c of an electromagnetic wave is determined by the constants of electricity and magnetism that is given as follow:

$$c = \frac{1}{\sqrt{\varepsilon_0\,\mu_0}} = 2.998 \times 10^8\,\frac{m}{s} \qquad (1\text{-}6)$$

It is considerable that there are still many puzzling aspects of the nature of light. Einstein wrote in 1951[7]: "All these fifty years of pondering have not brought me any closer to answering the question, what are light quanta?" The history of the photon started in 1901 with the formula by

---

[6] - Kevin Wray, An Introduction to String Theory, 2011, https://math.berkeley.edu/~kwray/papers/string_theory.pdf

[7] - L.B. Okun, Photon: history, mass, charge, 2008, https://arxiv.org/pdf/hep-ph/0602036.pdf

Planck for radiation of a black body. Planck published his article: "On the Law of the Energy Distribution in the Normal Spectrum" and wrote: "One can realize that the energy element" should be proportional to the number of vibrations v, so: $E = hv$, here h is the universal constants" [8].

In 1902, Lenard discovered that energy of electrons in photoeffect does not depend on the intensity of light, while it depends on the wavelength of the latter[9]. The history of theoretical efforts to define photon wave functions dates back to the early days of quantum mechanics and is still unfolding. However, there is not yet a consensus on the form a photon wave function should take or the properties it should have"[10].

In 1905, Albert Einstein suggested that electromagnetic waves could only exist as discrete wave-packets[11]. Gradually the Maxwell's equations were used as the basis for model of atoms[12].

The first time, in 1926, G. N. Lewis introduced the concept of "atom of light" that called "Photon"[13]. However, "A photon-like wave-packet based on novel solutions of Maxwell's equations is proposed. It is believed to be the first 'classical' model that contains so many of the accepted quantum

---

[8] - M. Planck, On the Law of the Energy Distribution in the Normal Spectrum, 1901, http://ffn.ub.es/luisnavarro/nuevo_maletin/Planck%20(1901),%20Energy%20distribution.pdf

[9] - Last reference L.B. Okun

[10] - Peter J. Mohr, Solutions of the Maxwell equations and photon wave functions, https://www.nist.gov/sites/default/files/documents/pml/div684/fcdc/photon-wave.pdf

[11] - Albert Einstein, German article "Über einen die Erzeugung und Verwandlung des Lichtes betreffenden heuristischen Gesichtspunkt", 1905, http://myweb.rz.uni-augsburg.de/~eckern/adp/history/einstein-papers/1905_17_132-148.pdf

[12] - Milan Perkovac, Maxwell's Equations as the Basis for Model of Atoms, Scientific Research, 2014, http://www.scirp.org/journal/PaperInformation.aspx?paperID=45334

Alexander Gersten, Maxwell Equations -The One-Photon Quantum Equation, Foundations of Physics , (2001), https://philpapers.org/rec/GERMEO

[13] - Helge Kragh, Photon: New light on an old name, 2014, https://arxiv.org/ftp/arxiv/papers/1401/1401.0293.pdf

features"[14]. So, to understand concept and properties of photon, we need to start with classical conceptions of electromagnetic radiation.

In classical mechanics, electromagnetic radiation is created when a charged particle is accelerated by an electric field, causing it to move[15]. The movement produces oscillating electric and magnetic fields, which travel at right angles to each other in a bundle of light energy called a photon.

Planck radiation formula was only a new scientific interpretation of classical electromagnetic theory. Because classical electromagnetic theory could not explain some of the new experiences such as photoelectric effect, Planck's radiation relation was accepted. Finally, physicists accepted the dual nature of light.

1.2.1   Particle-wave concept

The concept of "matter waves" or "de Broglie waves" reflects the wave-particle duality of matter. In 1923, Louis de Broglie, proposed a hypothesis to explain the theory of the atomic structure. By using a series of substitution de Broglie hypothesizes particles to hold properties of waves. Within a few years, de Broglie's hypothesis was tested by scientists shooting electrons and rays of lights through slits. What scientists discovered was the electron stream acted the same was as light proving de Broglie correct.[16]"

$$\lambda = \frac{h}{p} \text{ , } \lambda \text{ is de Broglie wavelength} \qquad (1\text{-}7)$$

Where, h is Plank' sconstant and p is momentum of a particle

---

[14] - John. E. Carroll, A photon-like wavepacket with quantised properties based on classical Maxwell's equations, 2006, https://arxiv.org/ftp/quant-ph/papers/0609/0609156.pdf

[15] - Jim Lucas, What Is Electromagnetic Radiation?, Live Science, 2015, https://www.livescience.com/38169-electromagnetism.html

[16] - Wave Nature of Electron, DeBroglie Wavelengths, Hyperphysics, http://hyperphysics.phy-astr.gsu.edu/hbase/debrog.html

De Broglie wavelength broke the boundary between particle and wave. Einstein said: "We are faced with a new kind of difficulty. We have two contradictory pictures of reality; separately neither of them fully explains the phenomena of light, but together they do "[17].

In quantum mechanics, the energy of the atoms could only take on discrete values, and these values depended on the frequency of the oscillation. But after emission the energy of photon is changeable, for example in the gravitational blueshift (or redshift) energy (and also frequency) of photon increases (or decreases). How we can explain the gravitational blueshift according to the relationship between photon energy and its frequency?

1.3 Mysteries zero rest mass and speed of the photon

Let's focus on energy and momentum of photon. After 1906 Einstein have derived the second postulate of special relativity the constancy of the speed of light by assuming that the light quanta that he proposed in 1905 were massless particles[18]. Relativistic energy and momentum is given by;

$$E = \frac{m_0 c^2}{\sqrt{1-\frac{v^2}{c^2}}} \quad and \quad P = \frac{m_0 v}{\sqrt{1-\frac{v^2}{c^2}}} \qquad (1\text{-}8)$$

It is just possible that we could allow $m_0 = 0$ provided the particle always travels at the speed of light[19] c. In this case above equations will not serve to define and so that for massless particle given by;

$$E = |P|c \qquad (1\text{-}9)$$

As it follows from the Einstein relativistic mass formula:

---

[17] - The world's first image of light as both a particle and a wave, ZME Science, 2015, http://www.zmescience.com/science/physics/light-particle-wave-03032015/

[18] - Field, J. Einstein and Planck on mass-energy equivalence in 1905-06: a modern perspective. 2014 arXiv preprint arXiv:1407.8507

[19] - Griffiths, D. Introduction to elementary particles: John Wiley & Sons, 2008, page 89

$$E^2 = m_0^2 c^4 + p^2 c^2 \qquad (1\text{-}10)$$

What does determine the momentum and energy of a massless particle? Not the mass (that is zero by assumption) not the speed (that is always c). Relativity offers no answer to this question, but curiously enough, quantum mechanics does, in the form of Plank's formula:

$$E = mc^2 = hv \Rightarrow m = \frac{hv}{c^2} \qquad (1\text{-}11)$$

Only moving photon has mass as follows from the Einstein formula $E = mc^2$. Physicists have not stopped on assumption of massless. There are more attempts were made to clarify the photon massless in theoretical and experimental physics. There are good theoretical reasons to believe that the photon mass should be exactly zero, there is no experimental proof of this belief[20]. These efforts show there is an upper bound on the photon mass, although the amount is very small, but not zero. The tight experimental upper bound of the photon mass restricts the kinematically allowed final states of photon decay to the lightest neutrino and/or particles beyond the Standard Model[21]. Theories and experiments have not limited to photons and graviton will also be included. For gravity, there have been vigorous debates about even the concept of graviton rest mass[22].

According to the general relativity theory, light moving through strong gravitational fields experiences a red- or blue shift. During the photon is falling in the gravitational field, its energy (mass) increases. According to $W = \Delta mc^2$, the force of gravity performs work on the photon, so the mass (energy) of the photon and its frequency increases (or decreases) from v to v' that given by;

$$v' = v(1 \pm \frac{GM_s}{rc^2}) \qquad (1\text{-}12)$$

---

[20] - Hojman, S. A., & Koch, B. (2013). Closing a window for massive photons. *Advances in High Energy Physics, 2013*

[21] - Heeck, J. How stable is the photon? *Physical review letters, 111*(2), 021801, (2013).

[22] - Goldhaber, A. S., & Nieto, M. M. Photon and graviton mass limits. *Reviews of Modern Physics, 82*(1), 939. (2010)

G is the gravitational constant; M is the mass of the body, c is the speed of light, r is the distance from the mass center of body. The plus sign refers to blueshift and minus sign refers to redshift. Also in the presence of gravity, the speed of light is not same for all observers. Einstein's derivation of the variable speed of light in a gravitational field potential[23] as follow:

$$c' = c(1 \pm \frac{GM_s}{rc^2}) \qquad (1\text{-}13)$$

Where c is the speed of light in vacuum and c' is the speed of light in gravitational field. It should be noted that there is no consensus about the speed of light in a gravitational field. For examples; so in the presence of gravity the speed of light becomes relative (variable depending on the reference frame of the observer). This does not mean that photons accelerate or decelerate; this is just gravity causing clocks to run slower and rulers to shrink[24]. The problem here comes from the fact that speed is a coordinate-dependent quantity, and is therefore somewhat ambiguous. To determine speed (distance moved/time taken) you must first choose some standards of distance and time, and different choices can give different answers. This is already true in special relativity: if you measure the speed of light in an accelerating reference frame[25], the answer will, in general, differ from c. Based on the Schwarzschild solution of the Einstein's equation of gravitational field, it is proved that the speed of light would change and the isotropy of light's speed would be violated in gravitational field with spherical symmetry[26].

---

[23]- To see the steps how Einstein theorized that the measured speed of light in a gravitational field is actually not a constant but rather a variable depending upon the reference frame of the observer: Einstein wrote this paper in 1911 in German:

http://www.physik.uni-augsburg.de/annalen/history/einstein-papers/1911_35_898-908.pdf

http://www.speed-light.info/speed_of_light_variable.htm

[24] - Variable Speed of Light, http://www.speed-light.info/speed_of_light_variable.htm

[25] - Philip Gibbs, Is The Speed of Light Constant? 1997,

http://www.desy.de/user/projects/Physics/Relativity/SpeedOfLight/speed_of_light.html

[26] - Mei Xiaochun, "A Test to Verify the Change of Light's Speed in the Gravitational Field of the Earth" https://arxiv.org/ftp/physics/papers/0611/0611294.pdf

In recent decades, the structure of photon[27] is discussed and physicists are studying the photon structure[28]. Some evidence shows the photon consists of a positive and a negative charges[29]. In addition, new experiment shows that the probability of absorption at each moment depends on the photon's shape[30], also photons are some 4 meters long which is incompatible with unstructured concept that means trying to describe the structure of photon is reasonable. So far, we know something about the photon, but on the properties of graviton almost nothing. General relativity does not help in this case, so we take the help of classical mechanics.

The problems caused by the lack of attention to the structure of photon and relationship between energy and frequency of it.

1.4. Gravitational effects on electromagnetic waves

The best way to recognize and understand electric and magnetic fields dependent on photon is reviewing the gravitational blueshift in a verified experiment. So a new look at behavior of photon in gravitational field and passing from limited and artificial barriers and boundaries of quantum mechanics and relativity can be useful to solve this riddle that a photon has made up from what particles.

If movements in the space are without any gravitational effects, photons move linearly with the speed of c (top of figure 1- 1). But space is full of gravitons. So, photon's paths are like the right side of figure (1-1).

Left side of figure (1- 1) shows that a photon is moving in a gravitational field of a massive body. In point A, the photon has the

---

[27] - Jonathan M. Butterworth, Structure of the Photon, Department of Physics and Astronomy, University College London

http://www.slac.stanford.edu/econf/C990809/docs/butterworth.pdf

[28] - Ilya Ginzburg, Physicists study photon structure, CERN COURIER, 1999,
http://cerncourier.com/cws/article/cern/28060

[29] - Hans W Giertz, The Photon consists of a Positive and a Negative Charge,

Measuring Gravity Waves reveals the Nature of Photons, 2013,

http://vixra.org/pdf/1302.0127v1.pdf

[30] - Victor Leong, et, at. Time-resolved Scattering of a Single Photon by a Single Atom, 2016,
https://arxiv.org/pdf/1604.08020v1.pdf

speed c, frequency v and energy E that reaches to point A. Gravitational field acts on the photon, some gravitons enter the structure of the photon. Photon accelerates toward the massive body. Its frequency, energy and speed increase. In point B, the photon has a frequency v_1, energy E_1 and speed of c_1. During the time that photon is falling, the distance between the the photon and body decreases, until it reaches to the point G. In point G frequency, speed and energy are maximum for this photon. When photon reaches point F', it is the same as point F, and so on. In point A', it is the the same as point A.

Path of light in space absence of gravity   *CPH Theory*

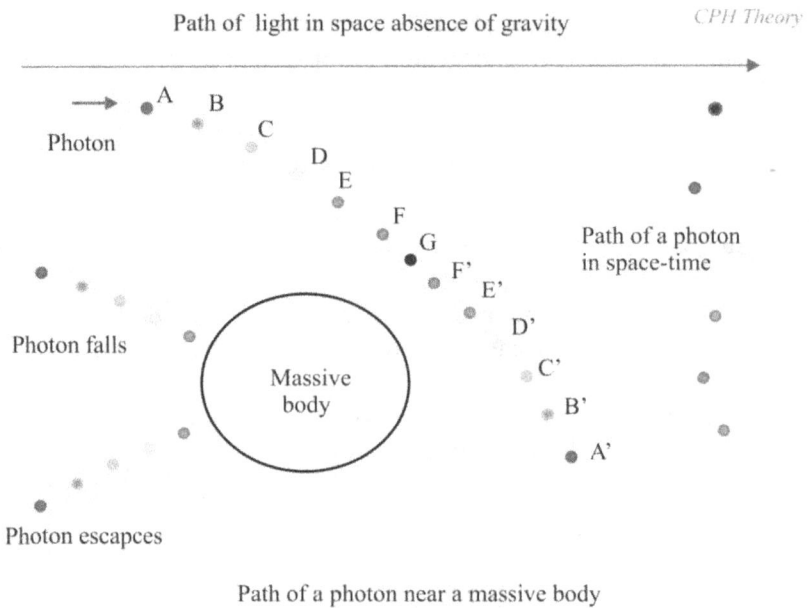

Path of a photon near a massive body

Fig 1-1: Paths of photons in space-time

The behavior of photons and gravitational fields is the same as spring and objects. On the left side of figure (1- 1), when a photon is falling, it shifts to blue and the gravity force converts into energy. When photon is escaping from a massive body it shifts to red and energy converts to gravity force.

1.4.1 The importance of attention to the structure of photon

Something that has been attractive in physics is behavior of light in different mediums and their interaction with other particles like electron. Doppler Effect, Photoelectric effect, Mossbauer Effect, curvature of space… all and all is analyzed without paying attention to structure of the photon. Something that has been pain attention and accepted by physicists in is that photon (and electron) is a point-like and unstructured particle. Point-like particles are mathematical abstractions with zero size. However, even zero-size particles have an extended effect, due to the effect of the field surrounding them[31].

The efforts and attempts to recognize and explain the structure of photon is an inevitable necessity. Due to this reason, CPH theory has formed based on a definition from the structure of photon. So how and where we can start to define the structure of photon? Such a definition must have both logical and experimental support, one of them for valid theories and another for experiments that are consistent with these theories. So, which theory can be helpful? Relativity or Quantum Mechanics?

Citing to both theories is valid, because both of these theories are famous among physicists. Fortunately, these theories have common experimental fields that citing to these common fields can help us to combine and unify both theories. Therefore, we continue to work by these common fields.

There is much evidence such as the Compton's effect, the pair production, the redshift and blueshift that lead us to accept photon has a structure. The changing frequency of the photon in the gravitational field has been demonstrated by the Pound-Rebka experiment[32]. The Pound–Rebka experiment is a well-known experiment to test Albert Einstein's theory of general relativity in 1959. The result confirmed the predictions of general relativity.

---

[31] - Don Lincoln, "What's the point?", Fermilab,
https://www.fnal.gov/pub/today/archive/archive_2013/today13-02-15_NutshellReadMore.html

[32]- R. V. Pound and G. A. Rebka, Jr., Gravitational Red-Shift in Nuclear Resonance, Phys. Rev. Lett. 3, 439, 1959,
https://journals.aps.org/prl/abstract/10.1103/PhysRevLett.3.439

1.4.2 The Pound–Rebka experiment

A photon with mass $m = \frac{hv}{c^2}$ has weight as $mg = \frac{hv}{c^2}g$, in gravitational field. When photon falls a distance equal y toward the earth, according to conservation law of energy we have:

$$hv' = hv + mgy = hv + \left(\frac{hv}{c^2}\right)gy \qquad (1\text{-}14)$$

$$v' = \left(1 + \frac{gy}{c^2}\right)v \qquad (1\text{-}15)$$

If we consider this phenomenon as another evidence to verify the general relativity, we will be stopped in the same old theories. Therefore, if we want to get a different result, we have to change our thoughts. The work that gravitational force does on the photon does not mean a simple concept of increasing in kinetic energy, but some deeper and more profound concepts are hidden beyond it. If we want to look at this phenomenon from the point of view of quantum field theory, we must accept that gravitons penetrate to the structure of the photon and in addition to the increasing its energy cause increasing electric and magnetic field intensity. Nevertheless, by considering the accepted concepts of quantum mechanics for gravitons, this phenomenon is not justifiable. Therefore, we need to reconsider the concepts of quantum mechanics about graviton and investigate about this phenomenon beyond quantum mechanics.

The Standard Model of particle physics describes the universe in terms of Matter (fermions) and Force (bosons). Particles of matter transfer discrete amounts of energy by exchanging bosons with each other[33]. So to generalize the relation between bosons and energy, we should start with gravity which is the weakest fundamental forces.

---

[33] - CERN, The Standard Model, https://home.cern/about/physics/standard-model

1.5 The discrete amount energy of graviton

Definition of the smallest discrete amount of energy is very vague and its detection is impossible. This ambiguity is due to reasonable restrictions based on experience, this is not just about physical limitations, even in mathematics we are dealing with some restrictions[34]. With all limits, behavior of photon in the gravitational field, helps we be able to define the smallest discrete amounts of energy. Consider a photon with energy $E = h\nu$ is escaping from a strong gravitational field. By reducing the frequency of photon (photon energy reduction), intensity of electric and magnetic fields are reduced too and finally, intensity of both the fields reaches to zero and the photon loses all its energy. Final limit for energy of photon before that reaches or tends to zero and still has spin, is equal to the smallest discrete amounts of energy that is given by:

Smallest discrete amount of energy $< h\nu, \forall \nu \ detectable$ (1- 16)

Regarding to gravity is the weakest fundamental forces which is transferring by graviton, relation(1-16) is defining the energy of graviton E_G that is given by:

$$E_G < h\nu, \forall \nu \ detectable \qquad (1\text{-} 17)$$

Where G is symbol of graviton. And the mass of the graviton m_G is given by:

$$m_G < \frac{h\nu}{c^2} , \forall \nu \ detectable \qquad (1\text{-} 18)$$

However, both of the above equations (1- 17 and 1- 18) do not show some intuitive value and assessable. In addition, the above equations do not have any specific information about electric and magnetic fields associated with photon. Therefore, we should be looking for an intuitive experience to be able to achieve tangible values and find out that how to produce electric and magnetic fields associated with the photon. Our approach for such this selection is changing the photon energy in a gravitational field that is

---

[34] - Finally, mathematicians accepted that there is three undefined terms point, line and plane in geometry.

https://www.reference.com/math/three-undefined-terms-geometry-97a228aecbb80a75

associated with the intensity of electric and magnetic fields of photon. It means that gravity works on the photon and gravitons enter to the structure of the photon in which it is justified according to the following equation:

$$F = -\frac{dU}{dx} \qquad (1\text{-}19)$$

Where dU changing potential energy at small distance dx and F is the force. Now we should explain the process of changing energy (equation 1-14) by using equation (1- 19). Photon falls at specified distance dy in gravitational field and its potential energy is reduced to dU (its kinetic energy increases). Reduction in potential energy of photon means that its frequency and kinetic energy increase that is given by dv and $dE = hdv$ respectively. From the perspective of quantum mechanics, that photon unstructured, it is impossible to explain this phenomenon. Therefore, if we want to analyze the structure of photon, we must pass the quantum space and enter into sub quantum space. From the perspective of sub quantum space, a number of gravitons enter into photon structure that is justified by equation (1- 19), and the energies of gravitons is added to the energy of photon. Now the question is how many gravitons enter into the structure of photon that generates the smallest possible change of energy that is given by $dE = hdv$? Also changing in energy of photon is associated with changing in intensity of electric and magnetic fields. Therefore, the work done on the photon by gravity must be discussed in such a way that justifies changing in intensity of electric and magnetic fields. Due to this reason in the CPH Theory, gravitons have properties that when gravity works on photon, can alter the intensity of electric and magnetic fields of the photon. This attitude led to the terms color-charge and magnetic-color in which they have used to define gravitons. In other words, identity of graviton changes without any change in its energy.

1.5.1 Considerable notes
Momentum of graviton: Graviton carries a discrete amount of energy and has momentum. The momentum equals mass times velocity, and velocity is a vector quantity while mass is a scalar quantity. A scalar multiplied by a vector is a vector. Remember relations (1- 5, 7, 9 and 1- 18). That is, graviton carries the force of gravity and it turns into energy. All forces

(fundamental and non-fundamental) are such, that is, force is converted into energy, and vice versa.

Particles and NR-particles: There are two kinds of particles in physics:

1-   Some particles like the photon moves only with the speed of light, in all inertial reference frames. Let's call these kinds of particles as Never at Rest condition particles (NR-particle).

2-   Other particles like the electron always move with the speed $v < c$ in all inertial reference frames, they could be called particles.

According to the above definitions, photon and graviton are NR-particle, while electron and proton are particles.

Concept of particle: Generally, we have almost the same understanding and imagination of large objects (at the level of molecules and larger). However, in the case of subatomic particles, there is no clearly defined and visualized concept, and there are many uncertainties, especially in the case of photon and graviton. Therefore, any theory offers certain understanding (such as loop and string theories) of these particles. So, in this book the particle is used, without any particular imagine of it.

1.6 Color-charges and magnetic-color

Now we are in a position that are able to take a new look at the structure of photon and define the features and properties of graviton in a way that is compatible with the feature of photon, and it is in accordance with experimental conditions very well. A photon with the lowest possible energy also carries electric and magnetic fields. Therefore, the features of gravitons entered into the structure of the photon must behave in a way that along with explaining the energy of photon, describes increasing in intensity of electric and magnetic fields. In other words, some of these gravitons cause increasing the electric field of photon and some other gravitons increase the intensity of magnetic field. Also, not only a photon at lowest level of its energy is formed by some of the gravitons, but also its formed members have electric and magnetic properties that is called color-charge and magnetic-color in CPH theory. The next step is to specify color-charges and magnetic-colors in which it is obtained by paying attention to at least change in energy of photon in a gravitational field (in a gravitational blueshift).

1.6.1 Formed elements of photon

Suppose a photon with NR-particle mass $m = hv/c^2$ and energy $E = hv$ falls from high h toward the earth relative to an inertial reference frame on the surface. Its frequency increases from v to v' , in fact, a number of gravitons enter into the structure of the photon such that $\Delta v = v' - v$. So the problem is; how many gravitons enter into the structure of photon to provide at least possible change of the energy of photon (minimum Δv)? So if Δv is minimum, then how many gravitons has entered into the structure of photon? What properties gravitons must have that they can be compatible with photons identity? For finding properties of gravitons and analyzing this process, we must observe these following conditions:

Condition1: photon is carrying two perpendicular electric field and magnetic field.

Condition2: The photon is electrically neutral and particles forming the electric field must neutralize each other.

Condition3: There are two groups, positive and negative color-charges in structure of photon that form photon's electric field and neutralize each other.

Condition4: Because these electric fields are moving, they create magnetic fields around themselves.

Condition5: simultaneously by producing positive and negative electric fields, two magnetic fields are produced around the electric fields do form. Therefore, it will be made two groups of magnetic-colors.

The above features necessitate that we consider each photon including four groups, two groups carry positive and negative electric effects and two groups carry magnetic effects. Suppose a photon with frequency v and energy hv is formed of n_1 elements, so that:

$$n_1 = n_{11} + n_{12} + n_{13} + n_{14}$$

Moreover, this photon with frequency v' and energy hv' is formed of n_2 elements, so that:

$$n_2 = n_{21} + n_{22} + n_{23} + n_{24}$$

For two levels of energy hv' , hv , we form the below matrices:

$$hv = \begin{bmatrix} n_{11} & n_{12} \\ n_{13} & n_{14} \end{bmatrix} \qquad (1\text{-}20)$$

$$hv' = \begin{bmatrix} n_{21} & n_{22} \\ n_{23} & n_{24} \end{bmatrix} \qquad (1\text{-}21)$$

Now, we consider the matrix of changing energy of photon $\Delta E = hv' - hv$ as follow:

$$\Delta E = \begin{bmatrix} A & B \\ C & D \end{bmatrix} \qquad (1\text{-}22)$$

Matrices (1- 20, 1- 21 and 1- 22) must satisfy the following equation:

$$hv' = \begin{bmatrix} A & B \\ C & D \end{bmatrix} + \begin{bmatrix} n_{11} & n_{12} \\ n_{13} & n_{14} \end{bmatrix} = \begin{bmatrix} n_{21} & n_{22} \\ n_{23} & n_{24} \end{bmatrix} \qquad (1\text{-}23)$$

According to conditions (1- 12), it will be determined elements A, B, C, D. We consider the first row of the matrix (1- 22), the elements of A, B for negative and positive colors charges. Element A represents positive color-charges and element B represents negative color-charges. In interaction between gravitons and photons, photon falls at specified distance dy and its energy increases (equation 1- 12) that due to equation (1- 19) the identity of a number of gravitons change by carrying gravitational force towards color-charges and enter to the structure of photon. We use the symbol of graviton G, for the both negative color-charge as G^- and positive color-charge as G^+, so that:

$$A = \kappa\, G^+ , \qquad B = \kappa\, G^-$$

Where κ is a natural number. In other words, when gravity works on photon, a number of gravitons enter into the structure of photon and photon's intensity of electric field increases, without any electrically effect and it is not created electric charge, because the photon is electrically neutral. So A, B must carry electric effect and their numbers must be equal (conditions2 and 3). Also according to the relative intensity electric fields and magnetic of electromagnetic waves $E = cB$, since color-charges and magnetic-colors are carrying electric and magnetic fields that they are NR-particles and countable, therefore, this relative can be replacement by a natural number such as κ, so we have $E = \kappa B$. When a number G^+ enter into photon structure, intensity of positive electric field of photon increases. Therefore, according to Maxwell's electromagnetic equations, the intensity of magnetic field increases, too. Therefore, the element C (equation1- 22) must increases the intensity of magnetic effect around the positive color-charges. Similarly, the element D must increases the intensity of magnetic field around the negative color-charges. The effect of these two elements are the same, but in terms of direction (which is proportional to the electric field) are different. Thus, according to the electric and magnetic field intensity we can be written:

$$C = G_m^+, \qquad D = G_m^-$$

The negative sign in relation $D = G_m^-$, only determines the direction of magnetic colors around the negative color-charges. So matrix (1- 22) that is called the CPH matrix, will be as follows:

$$CPH = \begin{bmatrix} \kappa\,G^+ & \kappa\,G^- \\ G_m^+ & G_m^- \end{bmatrix} \qquad (1\text{-}24)$$

According to the above expression, we are now able to define the least magnitude of a photon. A photon of minute energy contains some positive color-charges G^+, negative color-charges G^-, right rotation color-magnetic G_m^+ and left rotation color-magnetic G_m^- as shown in the CPH matrix (equation 1- 24). This very small energy can be express as the following;

Minute electromagnetic energy: $E_{Minute} = (2\kappa + 2)E_G$ (1- 25)

Thus, each photon is formed of a natural number E_{Minute}, so we have;

$$E = n(2\kappa + 2)E_G \text{ , or } \quad E = n\begin{bmatrix} \kappa\,G^+ & \kappa\,G^- \\ G_m^+ & G_m^- \end{bmatrix} \quad (1\text{-}26)$$

Equations (1- 17 and 1- 18) have previously defined the mass and energy of graviton. The longest wavelengths of radio production and broadcasting by radio stations is the world's longest wavelength which is more than 100 km[35]. Using the equation $m = h\nu/c^2$, the mass such photon is less than 10^{-40} kg. By comparing this value with the electron mass that is equal to 9.1×10^{-31} kg, and in the pair production a high-energy photon converts to an electron and a positron, each photon is containing billions and billions of gravitons.

1.7 Speed of graviton

In special relativity theory the speed of light in inertial frame of reference is constant c, and also it is the limit rate of speed. In general relativity, the speed of light is not constant (equation 1- 13). Reconsidering the principles of special relativity sometimes associated with new questions such as, whether is the constancy of the speed of light a natural law or is a natural accident? As the equation (1- 13) shows, whether there is no limit to speed of light in general relativity. Whether can the speed of light continue to infinity, like speed in classical mechanics?

The important concept in relationship between 'mass' and energy is c, regarding the phenomena of creation and decay of electron-positron pair, why do the related photons move at constant speed, but we could change the speed of matter and antimatter? What is the unique characteristic of matter, which is convertible to photons that move with constant speed c (speed of light)? The idea that object/particle could not travel at

---

[35] - Jim Lucas, What Are Radio Waves? Live Science, 2015, https://www.livescience.com/50399-radio-waves.html

superluminal speeds, originates from the structure of matter and the mechanism of interaction between field and mass; that with presenting a postulate we could generalize the constancy of speed from energy to mass and vice versa.

In gravitational blueshift, the energy of photon and consequently its frequency will increase. What is the mechanism of increasing in the photon energy that causes increase in its frequency? Are there more results than before in the energy-mass equivalence relation?

Photon is made up of color-charges and magnetic-color that have linear speed equal c with photon motion and nonlinear speed in the structure of photon, so they move faster than light speed. Therefore, the amount of passed path per unit of time is not equal c and it is greater than c, in the other word graviton moves faster than light speed. It is important that we note the speed of graviton (also color-charge and magnetic-color) that is given with V_G and as explained before, is faster than light speed, so $V_G > c$, that V_G is the total speed of transmission speed V_{GT} and non-transmission speed V_{GS} of graviton or color-charge and magnetic-color (figure 1- 2).

$$|V_G| = |V_{GT}| + |V_{GS}| > |c| \qquad (1\text{- }27)$$

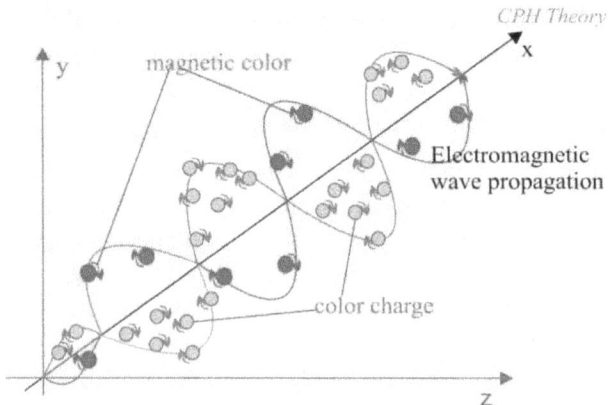

Fig1- 2; paths of gravitons in photon structure, color-charges and magnetic-color have spin and curvature speed

By presenting the graviton principle, the reverse of above process will be done to see how elementary particles like electrons and photons do form, and then we could describe the mechanism of the production of electromagnetic field, the strong and weak field.

1.8 Graviton Principle

Graviton is the most minuscule unit of energy with constant NR-particle mass m_G that moves with a constant magnitude of speed so that $|V_G| > |c|$, in all inertial reference frames. Any interaction between graviton and other existing particles represents a moment of inertia I where the magnitude of V_G remains constant and never changes. Therefore;

$$\nabla V_G = 0 \text{ , in all inertial reference frame and any space} \qquad (1\text{-}28)$$

Based on the principle of graviton, a graviton carries two types of energy generated by its movement in inertial reference frame, one is transmission energy E_{GT} and the other one is non-transmission energy E_{GS} , we can write:

$$E_G = E_{GT} + E_{GS} = constant \qquad (1\text{-}29)$$

As the graviton mass and speed is constant, its energy remains constant and can only its transmission energy changes to non-transmission energy and vice versa. Gravitons combine with each other and form energy and energy converts to matter and anti-matter and vice versa. In fact, everything has been formed of gravitons.

1.9   Sub-Quantum Energy (SQE)

We use CPH matrix (equation1- 24) to define sub quantum energy (SQE). The first column of CPH matrix is defined positive sub quantum energy SQE^+ and the second column of CPH matrix is defined negative sub quantum energy SQE^- , so;

Positive Sub Quantum Energy: $SQE^+ = \begin{bmatrix} \kappa G^+ \\ G_m^+ \end{bmatrix}$ (1- 30)

Negative Sub Quantum Energy: $SQE^- = \begin{bmatrix} \kappa G^- \\ G_m^- \end{bmatrix}$ (1- 31)

The amount of speed and energy of positive and negative sub quantum energies are equal, and the difference between SQE^+ and SQE^- are only in the sign of their color-charges and magnetic-color flow direction. So, sub quantum energy (regardless of sign, positive or negative) can be defined as follows:

Definition of sub quantum energy: Sub-quantum energy is the least electromagnetic energy that is defined as below:

$SQE = h\nu_{least,}$ $\nu_{least} < \nu, \forall E = h\nu,$ where $E = h\nu$ is detectable   (1- 32)

By comparing the equations (1- 17 and 1- 32), and definition of the sub quantum energy is determined $E_G < SQE$, and according to the equations (1- 26 and 1- 32) each photon is formed of the equal number of positive and negative sub quantum energies, and for even natural number n can be written:

$E = nSQE,$ where n is a natural number   (1- 33)

$E = nSQE = nm_{SQE}c^2 = n(m_{SQE}c)c = np_{SQE}c \Rightarrow E = np_{SQE}c$ (1- 34)

For two photons with energies E_1, E_2 we have;

$E_2 = h\nu_2 = n_2SQE, E_1 = h\nu_1 = n_1SQE, E_2 > E_1 \Rightarrow n_2 > n_1, n \propto \nu$ (1- 35)

Here, n_1 and n_2 are natural numbers. With increasing a photon's energy, its frequency also increases. Thus there should be a logical explanation between energy increasing and frequency increasing. Therefore, based on SQE definition and equation (1- 35) we can relate the relation between photon's energy and frequency and the interaction between SQEs in a photon's structure, i.e. with increasing the number of SQEs in photons, the interaction between SQEs in photons will increase and the frequency that originates from the interaction between SQEs will increase, too.

Note: Although $n \propto \nu$, this proportion does not necessarily represent an equation, but simply represents the physical fact that frequency has direct relation with the number and interaction of *SQEs* in a photon. Besides the relation between *SQEs* and ν, could conclude that the linear speed of *SQE* in a vacuum relative to the inertial frames of reference, is actually the speed of light c. Since *SQE* in a photon's structure has a linear speed equal to c and it has nonlinear motions, the real speed of *SQE* is when all *SQE* nonlinear motions turn into linear motion and it only takes linear motion. In other words, the limit speed of *SQE* is V_{SQE} which is faster than light speed c, i.e. $|V_{SQE}| > |c|$. Consider that in special relativity the light speed is constant, and in general relativity besides increasing of photon frequency while falling in a gravitational field, its speed also increases (equation 1-13); that we could take it as a proof of $|V_{SQE}| > |c|$.

$$|V_G| > |V_{SQE}| > |c| \qquad (1\text{-}36)$$

1.10 Sub-Quantum Energy Principle

One *SQE* is a very small energy with *NR-particle* mass m_{SQE} that moves with speed greater than c, $|V_{SQE}| > |c|$ relative to inertial reference frame and in every interaction between *SQEs* with other particles or fields the amount speed of *SQE* remains constant; as in every physical condition we have;

$$\nabla V_{SQE} = 0 \text{ , in all inertial reference frames and any space} \qquad (1\text{-}37)$$

SQE principle (equation 1- 37) shows that in every condition the mass, energy and the amount speed of *SQE* remains constant, and only the transmission speed V_{SQET} and energy E_{SQET} of *SQE* convert to its non-transmission speed V_{SQES} and energy E_{SQES} , and vice versa. So, we have;

$$|V_{SQE}| = |V_{SQET}| + |V_{SQES}| = constant \qquad (1\text{-}38)$$

$$|E_{SQE}| = |E_{SQET}| + |E_{SQES}| = constant \qquad (1\text{-}39)$$

1.11 Virtual Photon & Quantum Vacuum Energy

In contrast to classical mechanics, in quantum mechanics, the vacuum is not empty and it has energy. A quantum fluctuation is the temporary change in the amount of energy in a point in space, as explained in Werner Heisenberg's uncertainty principle[36]. Heisenberg uncertainty equation is obtained by multiplying the energy uncertainty ΔE and time uncertainty Δt is always greater than zero[37] that is given as follow;

$$\Delta E \cdot \Delta t \geq \frac{\hbar}{2} \qquad (1\text{-}40)$$

Where $\hbar = \frac{h}{2\pi}$ and h is Planck constant. Based on quantum field theory, the quantum vacuum is filled with virtual particles which are in a continuous state of fluctuation. Virtual particle-antiparticle pairs are created from vacuum and annihilated back to it. These virtual particles exist for a short time dictated by Heisenberg uncertainty relation[38]. How we can explain the zero-point energy without using the uncertainty principle?

Following the success of the theory of relativity and quantum mechanics, the theory of relativistic charged particles interacting with electromagnetic fields was formulated.

In the perturbative approach to quantum field theory, the full field interaction terms are approximated as a perturbative expansion in the number of particles involved. Each term in the expansion can be thought of as forces between particles being mediated by other particles. In QED, the electromagnetic force between two electrons is caused by an exchange of photons. Richard Feynman devised a short hand way of writing out particle

---

[36] - Brandon West, "Vacuum Energy: Proof of Free Energy in the Space All Around Us" WakingTimes, 2014, http://www.wakingtimes.com/2014/04/11/vacuum-energy-proof-free-energy-space-around-us/

[37] - Uncertainty and Virtual Particles, https://pdg.web.cern.ch/pdg/cpep/unc_vir.html

[38] - H. Razmi and S. M. Shirazi, "Is the Free Vacuum Energy Infinite?" https://arxiv.org/ftp/arxiv/papers/1302/1302.1433.pdf

interactions called Feynman Diagrams[39]. Under QED, charged particles interact by the exchange of virtual photons, photons that do not exist outside of the interaction and only serve as carriers of momentum/force. In fact, quantum field theory (QFT) is the mathematical and conceptual framework for contemporary elementary particle physics[40], regardless to how fermions produce bosons and what the mechanism of exchanging bosons is.

These ambiguities in mathematics calculation are due to lack of attention to the structure of photon and mode of electromagnetic energy production. Is there a way to explain virtual photon (in fact interaction between charged particles) without using the uncertainly principle? Is there a relationship between quantum vacuum energy processes and the interactions of charged particles? In quantum electrodynamics (QED) a charged particle emits exchange force particles continuously. This process has no effect on the properties of a charged particle such as its mass and charge. How is it explainable? If a charged particle as a generator has an output known as a virtual photon, what will be its input?

If we replace the force by transferring energy - momentum, all physical interactions are justifiable. Moreover, without using force, we can describe all physical processes and interactions. If we use force for simplify of mathematical relations, we should not forget that there is physical concepts and quantities beyond these mathematical and undeniable (which is explained in this book).

In fact, by describing the quantum vacuum energy and generalization the Maxwell's equations of electromagnetism to gravity, we can describe the mechanism of quantum vacuum energy and the production of electric fields process by charged particles.

---

[39] - FLIP TANEDO, Let's draw Feynman diagrams! Quantum Diaries, http://www.quantumdiaries.org/2010/02/14/lets-draw-feynman-diagams/

[40] - Meinard Kuhlmann, quantum Field Theory, The Stanford Encyclopedia of Philosophy (2012), https://philpapers.org/rec/KUHQFT

1.11.1 Virtual Photon from Sub-Quantum Point of View

When a photon falls in a gravitational field as Δr, the graviton's density in the vicinity of the photon electric field changes the value of ∂G_E, because the intensity of electric field changes as E_G (E is the electric field arising from gravitons equations (1- 24 to 1- 26). In fact gravitons enter the structure of photon, and the intensity of electrical and magnetic fields which depends on photon increases. Two types of gravitons should enter the photon structure, so that they are able to increase the intensity of photon electric field without any charge effect. Thus the interaction between gravitons and photon, negative and positive G^-, G^+ gravitons are produced and enter the photon structure. The photon moves in the same direction as the increasing intensity of the gravitational field does, and the photon electric field is perpendicular to the photon movement direction that is compatible with the following equation:

$$\nabla \times E_G = -\frac{\partial G_E}{\partial t} \Leftrightarrow i(G^+, G^-) \qquad (1\text{-}41)$$

By changing the photon electric field, magnetic field also changes. In this case also, the gravitons are converted into magnetic carrier particles G_m^+, G_m^- and enter the structure of photon that is given by;

$$\nabla \times B_G = \mu_0 \varepsilon_0 \frac{\partial E_G}{\partial t} \Leftrightarrow j(G_m^+, G_m^-) \qquad (1\text{-}42)$$

Where i and j are natural numbers, and proportion between i and j should be consistent with equation (1- 26). Assume $2k$ positive and negative color-charges (kG^+, kG^-) enter to a very small part of photon structure, proportional to the number of color-charges, the number of magnetic-colors are produced around the color-charges. Two opposite electric field are created in this space. Around each of the electric field a magnetic field is created by magnetic-colors. According to the sign of the electric fields, direction of magnetic fields are different, each magnetic field cover its color-charges and prevents them of escaping (equations 1- 30 and 1- 31). Each of the magnetic fields protects its electric field and prevents them from collapsing. This mechanism is justifiable by Larmor radius (gyro radius or radius of the cyclotron)[41] that is given by the following equation:

$$r_g = \frac{mv_\perp}{|q|\mathbf{B}}$$ (1- 43)

Where r_g is the gyro radius, m is the mass of the charged particle, $v_\perp$ is the velocity component perpendicular to the direction of the magnetic field, q is the charge of the particle, and $\mathbf{B}$ is the constant magnetic field. This defines the radius of circular motion of a charged particle in the presence of a uniform magnetic field.

1.11.2 Positive and negative sub quantum energies

When density of color-charges changes in the structure of a photon, then the density of magnetic-color changes, too. Therefore, the electric fields do not decay in the structure of photon. In general, we are able to describe the sub quantum energy (SQE), virtual photons and real photon as follows:

1- **Positive Sub Quantum Energy (SQE^+):** The positive sub quantum energy (equation 1- 30) is a set of positive color-charges with its affiliates magnetic-color that is shown by right wedge $\triangleright$ (the first column of the CPH matrix, equation 1- 24), which is defined as follows:

$$\text{Positive Sub Quantum Energy } SQE^+ : \triangleright= \begin{bmatrix} \kappa G^+ \\ G_m^+ \end{bmatrix}$$ (1-44)

2- **Negative Sub Quantum Energy (SQE^-):** The negative sub quantum energy (equation 1- 31) is a set of negative color-charges with its affiliates magnetic-color that is shown by left wedge $\triangleleft$ (the second column of the CPH matrix, equation 1- 24), which is defined as follows:

$$\text{Negative Sub Quantum Energy } SQE^- : \triangleleft= \begin{bmatrix} \kappa G^- \\ G_m^- \end{bmatrix}$$ (1- 45)

---

[41] - Francis F, & Chen, "*Introduction to Plasma Physics and Controlled Fusion*", Plenum Press. 1984

1.11.3 Photon and virtual photons

Virtual photons: There are two types of virtual photons, positive virtual photon γ^+ and negative virtual photon γ^- that each of them is formed of number same-sign sub quantum energies, which is defined as follows:

$$\text{Positive virtual photon; } k \rhd = \gamma^+ \qquad (1\text{-}46)$$

$$\text{Negative virtual photon; } k \lhd = \gamma^- \qquad (1\text{-}47)$$

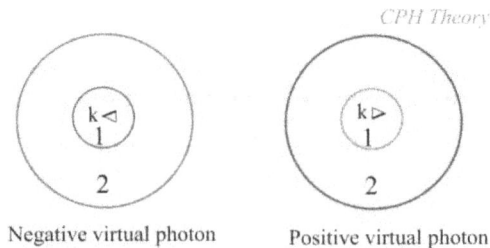

Negative virtual photon   Positive virtual photon

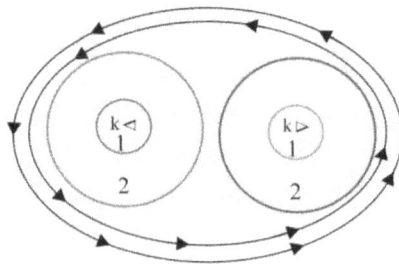

A photon is formed of $k \rhd + k \lhd$, but magnetic fields
around $k \rhd$ and $k \lhd$ prevent them from combination

Fig1- 3; structure of photon

Photon: A real photon is formed of a positive virtual photon and a negative virtual photon:

$$\gamma^+ + \gamma^- = \gamma \qquad (1\text{-}48)$$

$$(n \rhd + n \lhd) = n(\rhd + \lhd) \ or \ n|\rhd\rangle + n|\lhd\rangle = \gamma \qquad (1\text{-}49)$$

Where, $n \ and \ k$ are natural numbers. So far, the production of electromagnetic energy (photons) was described by using gravitational blueshift, in reverse phenomena photons decay to negative and positive virtual photons.

In gravitational redshift, virtual photons also decay to positive and negative sub quantum energies (SQEs), and sub quantum energies (SQEs) decay to color-charges and magnetic-colors, too. Color-charges and magnetic-colors away from each other, lose their effect on each other and become gravitons (figure 1- 3). So, photons are combination of positive and negative virtual photons. Photon is a very weak electric dipole that is consistent with the experience. In addition, this property of photon (very weak electric dipole) can describe the absorption and emission energy by charged particles.

1.12 Light Speed

According to the principle of special relativity, the speed of light in vacuum is constant and it equals to c for all inertia observers, and it is independent of the light source. How we can conclude this principle by using sub quantum energy principle? First, according to principle of SQE (which is also the result of the graviton principle) the amount of the linear speed of SQE depends to the interaction between $SQEs$ and the other particles (or fields) in the medium. So, in a vacuum, photon (light) has not any interaction with other particles or fields outside of the photon structure, (assume gravitational effect of vacuum is negligible), thus, the linear speed of $SQEs$ in the structure of photons are constant and equal to $V_{SQET} = c$. Also, the linear speed of virtual photons in a vacuum is the same amount of c. Let's in generally, show the speed of photons as v_{light}, it changes from one medium to another one and in a vacuum is c, it means the speed of light in vacuum also is $v_{light} = c$. So that:

$$\nabla v_{light} = 0 \qquad (1\text{-} 50)$$

Thus, the linear speed of photon depends to medium conditions. Same as gravitons and sub quantum energy, but the total amount of transmission speed v_{lightT} and non-transmission speed v_{lightS} of photon is constant and it is equal to $|v_{light}|$, by changing the medium conditions, such as photon enters to water, a part of its linear speed converts to non-linear speed and in this case we have $v_{lightT} < c$. But this effect is not permanent, because the medium conditions (with any physical and chemical conditions) cannot affect the amount speed of v_{light}, and medium only can temporarily changes linear motion to non-linear motion and vice versa. That is why as soon as the light comes out of a medium; the new medium will be affected on it. So we can write:

$$|v_{light}| = |v_{lightT}| + |v_{lightS}| = constant \quad (1\text{-}51)$$

As the principle of sub quantum energy shows, the total transmission speed and non-transmission speed of SQE is always constant relative to the inertial reference frame and it is an intrinsic property of nature, which is also affected by the graviton principle, because SQE of gravitons are made. So the amount transmission speed (in this case linear velocity) of SQE is independent of emitter source of light. On the other hand, photon is made up of positive and negative $SQEs$ (equation 1- 33), so the linear speed of photon depends to linear speed of $SQEs$ that in vacuum is c *(like the speed of light in water or any other medium)*. But in the liquid (e.g. water) it is less than c, and under the influence of the gravitational field, according the direction of gravity force is applied, this amount varies and it can be more or less than the speed of light in a vacuum (equation 1-13).

So the speed of light in a vacuum is constant and it is equal to c for all the inertia observers, but in accelerating frame, as well as Einstein's General Relativity has made clear, is not constant (equation 1-13). Now the question arises that how much change will increase or decrease the speed of light? The answer to this question is given later (see singularity section).

1.13 Zero Point Energy (ZPE)

The Casimir effect is a small attractive force that acts between two close parallel uncharged conducting plates. The effect was predicted by Casimir in 1948. The pressure on the plates due to vacuum fluctuations differs on the inside faces, from that on the outside, as long wavelength modes are excluded. The difference in these pressures produces the Casmir force[42]. According to Casimir, Bohr mumbled something about zero-point energy being relevant[43] (figure 1-4).

To describe zero point energy we should consider that space is full of gravitons and graviton is a basic element to produce energy. There is no physical point in space devoid of gravity effect. Therefore, at any physical point in the space, there are energy production facilities.

Under the terms of *SQE*, any space that has the gravitational effects can produce electromagnetic energy, and here the photon in the conversion of gravitons into G^-, G^+, G_m^-, G_m^+ , and electromagnetic energy acts only as a catalyst. When intensity of gravitational field increases or interfere gravitational fields of two massive bodies that are moving adjacent each other, gravity produces the electromagnetic energy. But the amount of electromagnetic energy in space depends to density of gravitons $\rho(G)$ in the space. Therefore, the integral on space compared to the density of gravitons, namely:

$$E = \oiiint \rho(G)dV = n(\kappa G^+, \kappa G^-, G_m^+, G_m^-) \qquad (1\text{-}52)$$

Where V is volume. According to equations (1- 43 to 1- 48) we can describe the mechanisms of zero-point energy production.

When the density of graviton increases in space, a number of gravitons with the NR-particle mass (m_G) are adjacent to each other and interactions

---

[42] - Casimir effect, Philosophy of Cosmology,
http://philosophy-of-cosmology.ox.ac.uk/casimir-effect.html
[43] - Calphysics, Zero-Point Energy, http://www.calphysics.org/zpe.html

are logged and they are converted to color-charges and a number gravitons convert to magnetic-color.

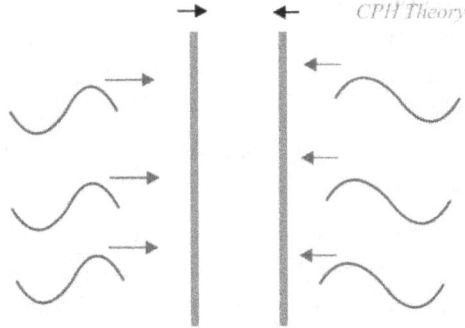

Fig1-4; the difference pressures between inside faces and outside of plates causes the Casmir force

Finally, sub quantum energies produce virtual photons, and virtual photons form the real photon. About the vacuum energy, even in the absence the photons in vacuum, the Maxwell's equations can be generalized in vacuum. Equation (1- 41) comes in just below, but the equation (1- 42) remains as before.

$$\nabla \times G_E = -\frac{\partial G}{\partial t} \Leftrightarrow i(G^+, G^-) \qquad (1\text{-}53)$$

In equation (1- 53), ∂G shows even without the electric field, when the density of graviton increases in space, gravitons interacting with each other and they acquire electric field and magnetic and they produce the electromagnetism energy. According to the above description and with regard to the phenomenon of gravitational redshift and blueshift, in general it can be concluded that:

Gravitational energy $\Leftrightarrow$ Electromagnetic energy   (1- 54)

If we reconsider the gravitational redshift of a black hole, photon loses all its energy in escaping and eventually reaches to $v' = 0$ (equation 1- 12), and there is no energy in photon structure to escape of black hole. In fact, all the photon's energy is converted into gravitons.

1.13.1 Hawking Radiation

In a simplified version of the explanation, Hawking predicted that energy fluctuations from the vacuum cause the generation of particle-antiparticle pairs near the event horizon of a black hole. One of the particles falls into the black hole while the other escapes before they have an opportunity to annihilate each other[44]. The net result is that to someone viewing the black hole, it would appear that a particle has been emitted.

How is Hawking radiation explainable by the equation (1- 52) for the ZPE? To resolve this problem, there are three aspects of a black hole to be considered;

•The density of gravitons is extremely high around a black hole.

• Gravitons convert to photons rapidly.

• The Dirac equation shows how photons produce matter and anti-matter.

According to the above expression, the space around a black hole produces high energy photons whose energy is enough for pair production.

In a black hole situation, n becomes large, so, E (equation 1- 52) is comparable to the total mass of a particle and anti-particle. This process does not need take into account the time factor that the uncertainty principle dictates in equation (1- 40).

So, pair production is a common occurrence around a black hole. Yet even without heavy nuclei collide to produce a pair of photons - photons[45].

---

[44] - BLACK HOLE THEORY & HAWKING RADIATION, The Physics of the Universe
http://www.physicsoftheuniverse.com/topics_blackholes_theory.html

[45] - Robert J. Gouid, "Pair Production in Photon-Photon Collision", Physical Review, Volume 155, Number 5, 1967,

1.14 Spin of Graviton

In standard model, the graviton is a massless particle and it must have a spin of 2 which moves with the speed of light. "We cannot solve problems by using the same kind of thinking we used when we created them." Einstein said[46]. To solve old problems, new ideas are needed. In the previous parts we have discussed about the speed of graviton and it was shown that the graviton should have speed greater than the speed of light.

It was explained that the graviton is a NR-particle, so it is inappropriate to speak about rest mass of graviton. But about the graviton spin, we should be cautious about the spin of the graviton and not a final verdict, because there is difference between a free graviton (in the far distance of objects and particles) and a graviton that is interacting with other particles.

Pair production and decay shows that a photon with spin one converts to two particles (electron and positron) with spin $1/2$. In quantum physics, it usually is used the new observing phenomena for confirming valid theories, only - in this case the pair production verification E and Dirac equation - and not think beyond that. The amount speed and energy of graviton is constant (equations 1- 28 and 1- 29), but by changing the amount transmission energy and speed of graviton, also altered the identity of its spin, without changing the total amount of its speed and energy.

---

http://fcaglp.fcaglp.unlp.edu.ar/astrofrelat/astrofisica/media/Papers/Gould_Schreder_196 6.pdf

[46] - Albert Einstein Quotes, https://www.brainyquote.com/quotes/quotes/a/alberteins121993.html

2

Sub quantum electrodynamics

Newton's second law is motion equation in classic mechanics that does not say anything about the nature of force. The equivalent formulations and their extensions such as Lagrangian and Hamiltonian do not explain about mechanism of converting Potential energy to Kinetic energy and Vice versa. In quantum mechanics, Schrodinger equation is similar to Newton's second law in classic mechanics. Quantum mechanics is also extension of Newtonian mechanics to atomic and subatomic scales and relativistic mechanics is extension of Newtonian mechanics to high velocities near to velocity of light too. Schrodinger equation is not a relativistic equation, because it is not invariant under Lorentz transformations. Dirac expanded The Schrodinger equation by presenting Dirac Sea and founded relativistic quantum mechanics.

In this section by reconsidering the Dirac Sea and his equation, the structure of photon is investigated. Then according to structure of photon and charged particles sub quantum electrodynamics is described.

2.1 Klein-Gordon Equation

The Klein-Gordon equation is the first important step from non-relativistic quantum mechanics towards relativistic quantum mechanics. The Klein-

Gordon equation is a relativistic version of Schrodinger equation[1] that was presented as follows[2]:

$$E^2 = p^2c^2 + (mc^2)^2 \qquad (2\text{-}1)$$

In quantum mechanics, momentum of a particle P (to plane wave, to wave vector) is given as $P = \hbar k$ in which k is wavenumber and $\hbar = \dfrac{h}{2\pi}$. Moreover, a particle with energy E has the frequency ω that is indicated by relation $E = \hbar\omega$. So by interpreting quantum mechanics operators, we can write:

$$P \rightarrow -i\hbar\nabla \ , \ E \rightarrow i\hbar\frac{\partial}{\partial t} \qquad (2\text{-}2)$$

Then the relativistic form of Klein-Gordon Equation will be expressed as follows:

$$\frac{1}{c^2}\frac{\partial^2}{\partial t^2}\psi - \nabla^2\psi + \frac{m^2c^2}{\hbar^2}\psi = 0 \qquad (2\text{-}3)$$

The solutions of this equation is complex values of wave function $\psi(t,x)$. By taking radical from both sides of relation (2- 1), we have:

$$E = \pm\sqrt{p^2c^2 + (mc^2)^2} \qquad (2\text{-}4)$$

It is natural that we try to use relativistic form of Klein-Gordon equation by using the nature of energy in special relativity (in relation 2- 1), so combining the relations (2- 1, 2- 2 and 2- 3) and neglecting from negative part of relation (2- 4) (because the negative energy is meaningless), we have:

$$E = \sqrt{p^2c^2 + (mc^2)^2} \qquad (2\text{-}5)$$

---

[1] - Schrödinger's equation — what is it? https://plus.maths.org/content/schrodinger-1
[2] - Relativistic Quantum Mechanics, http://hitoshi.berkeley.edu/221B-S02/Dirac.pdf
The Klein-Gordon Equation, http://www.mysearch.org.uk/website1/html/538.Klein-Gordon.html

Then, by putting just mechanic quantum operators for momentum and energy in relation (2- 5), we will have the following equation:

$$\sqrt{m^2c^4 - \hbar^2\nabla^2c^2}\,\psi = i\hbar\frac{\partial}{\partial t}\psi \qquad (2\text{- }6)$$

In relation (2- 6), differential operator ∇ lies under radical that is meaningless. If we expand under the radical (left hand side of relation 2-6), we will have[3]:

$$\sqrt{m^2c^4 - \hbar^2\nabla^2c^2} = mc^2\sqrt{1 - \left(\frac{\hbar}{mc}\nabla\right)^2}$$
$$= mc^2\left[1 - \frac{1}{2}\left(\frac{\hbar}{mc}\nabla\right)^2 + \frac{1}{8}\left(\frac{\hbar}{mc}\nabla\right)^4 + \cdots\right]$$

By neglecting from third term onwards, we will have:

$$i\hbar\frac{\partial}{\partial t}\psi = mc^2\left[1 - \frac{1}{2}\left(\frac{\hbar}{mc}\nabla\right)^2\right] = \left(mc^2 - \frac{\hbar^2}{2m}\nabla^2\right)\psi \qquad (2\text{- }7)$$

$$i\hbar\frac{\partial}{\partial t}\psi = \left(mc^2 - \frac{\hbar^2}{2m}\nabla^2\right)\psi \qquad (2\text{- }8)$$

Equation (2- 8) is the same Schrodinger equation in which the first term in right hand side is the zero rest energy of a particle[4]. It is obviously seen that this equation is not invariant under Lorentz transformations, because in this equation, derivative with respect to time is first degree and with respect to place is second degree. Therefore, it is non-relativistic. Moreover, it was neglected from negative section of relation (2- 4) due to not being acceptable of negative energy. But Dirac did not ignore existence of negative energy.

[3] - Tobias Gleim, "Klein-Gordon and square-root operator equations for two-spinors and scalars: perturbation calculations for hydrogen-like systems" https://arxiv.org/ftp/quant-ph/papers/0602/0602047.pdf

[4] - The Klein-Gordon equation, https://www.eng.fsu.edu/~dommelen/quantum/style_a/kg.html

2.2 Dirac equation

In 1928, Paul Dirac published a paper entitled "The Quantum Theory of the Electron" that presented relativistic form of wave equation for electron in which it became the main instruction for obtaining Dirac equation[5]. Dirac equation is generalization of Schrodinger equation to compute wave function of particles that is consistent with special relativity too. Dirac extended this equation based on Klein-Gordon equation that had efficiency in interpretation of states with negative energy, it means that it covered negative part of equation (2- 4). Therefore; Dirac presented his equation as follows:

$$[p_0 + \rho_1(\sigma, P) + \rho_3 mc]\psi = 0 \qquad (2\text{-} 9)$$

In which ρ_1, ρ_3 has been taken from Pauli matrices[6]. Dirac equation justifies wave function of particles with half integer spin like fermions (the same as electron), while Klein-Gordon equation is considered for particles with spin of zero (like certain mesons). Dirac also could predict existence of anti-matter with his equation that later it was verified with experiment too. 30 years later in 1958, Dirac suggested the main form of his equation by publishing a book[7] as follows:

$$\left(\beta mc^2 + \sum_{j=1}^{3} \alpha_j\, cp_j\right)\psi(x,t) = i\hbar \frac{\partial \psi(x,t)}{\partial t} \qquad (1\text{-} 10)$$

In which $\psi(x, t)$ is a wave function for an electron with rest mass m with space-time coordinates x, t. The elements $p_1,\ p_2,\ p_3$ are coordinates of

---

[5] - Dirac P. A. M., "The Quantum Theory of the Electron", Proceedings of the Royal Society of London. Series A, Containing Papers of a Mathematical and Physical Character 117 (778): 610–24. , Published 1 February 1928

http://rspa.royalsocietypublishing.org/content/royprsa/117/778/610.full.pdf

[6] - Mario Bacelar Valente, "The Dirac equation, the concept of quanta, and the description of interactions in quantum electrodynamics", http://philsci-archive.pitt.edu/8366/1/The_Dirac_equation,_the_concept_of_quanta,_and_the_description_of_interactions_in_quantum_electrodynamics.pdf

[7] - Paul Dirac, The Principles of Quantum Mechanics, 1958, Oxford University Press. p. 255. ISBN 978-0-19-852011-5, Wikipedia, https://en.wikipedia.org/wiki/Dirac_equation#cite_note-3

momentum that are recognized as momentum operators in Schrodinger equation. There basic physical constants reflect properties and virtues of relativity and quantum mechanics.

The smallest representation of a_i and β is as 4×4 matrices and they can be produced by using Pauli matrices σ_i as sub matrices as follows[8]:

$$\beta = \begin{pmatrix} 1 & 0 \\ 0 & -1 \end{pmatrix} \quad \sigma_i = \begin{pmatrix} 0 & \sigma_i \\ \sigma_i & 0 \end{pmatrix} \quad (2\text{-}11)$$

Such that each element is a 2×2 matrix that can completely written as follows:

$$\beta = \begin{bmatrix} 1 & 0 & 0 & 0 \\ 0 & 1 & 0 & 0 \\ 0 & 0 & -1 & 0 \\ 0 & 0 & 0 & -1 \end{bmatrix} \quad (2\text{-}12) \; \sigma_1 = \begin{bmatrix} 0 & 0 & 0 & 1 \\ 0 & 0 & 1 & 0 \\ 0 & 1 & 0 & 0 \\ 1 & 0 & 0 & 0 \end{bmatrix}$$

$$\sigma_2 = \begin{bmatrix} 0 & 0 & 0 & -i \\ 0 & 0 & i & 0 \\ 0 & -i & 0 & 0 \\ i & 0 & 0 & 0 \end{bmatrix} \quad \sigma_3 = \begin{bmatrix} 0 & 0 & 1 & 0 \\ 0 & 0 & 0 & -1 \\ 1 & 0 & 0 & 0 \\ 0 & -1 & 0 & 0 \end{bmatrix} \quad (2\text{-}13)$$

Now, by combining equations (2- 1 and 2- 10), it can be investigated difficulty of explaining negative energy by a different approach[9]:

$$E^2 = p^2 c^2 + (mc^2)^2 = \left(\beta mc^2 + \sum_{j=1}^{3} \alpha_j \, cp_j\right)^2 \quad (2\text{-}14)$$

For a particle in special case $p = 0$, we will have:

$$E^2 = (mc^2)^2 = (\beta mc^2)^2 \quad (2\text{-}15)$$

By considering β matrix (relation 2- 12), we can write:

[8] - Quantum Theory 2015/16 , http://www2.ph.ed.ac.uk/~bjp/qt/rqt.pdf

[9] - ANDREW ERIC BRAINERD, "SELF-ADJOINT EXTENSIONS TO THE DIRAC COULOMB HAMILTONIAN" MASSACHUSETTS INSTITUTE OF TECHNOLOGY, 2010, https://dspace.mit.edu/bitstream/handle/1721.1/61203/701106459-MIT.pdf;sequence=2

$$\beta mc^2 \rightarrow \begin{bmatrix} mc^2 & 0 & 0 & 0 \\ 0 & mc^2 & 0 & 0 \\ 0 & 0 & -mc^2 & 0 \\ 0 & 0 & 0 & -mc^2 \end{bmatrix} \quad (2\text{-}16)$$

For eigenvalues and considering $p = 0$ (in equation 2- 4), we will have[10]:

$$E_+ = mc^2, \ E_- = -mc^2 \quad (2\text{-}17)$$

Dirac equation predicted existence of a particle with negative energy and he was confronted with unbelievers of physicists. However, in 1932, Anderson[11] discovered this particle in cosmic ray and they called it "positron". Later, pair "electron-positron" was created in the laboratory by photon decay process.

A photon with high energy loses all its energy $E = h\nu$ in collision with nucleus and creates pair "electron-positron". Positron is a particle that has all the same properties of electron except in electric charge and the sign of its magnetic moment. Because electric charge of positron is positive. Existence of negative energy in Dirac equation was not pleasant for physicists. Nevertheless, negative energy in this equation caused that Dirac discussed on negative energy in general and published it through a paper in 1930 (before discovery of positron) entitled: "Theory of Electrons and Protons"[12].

2.3 Dirac Sea

Dirac Sea is a theoretical model that introduces vacuum as a sea of infinite particles with negative energy. Dirac presented this model in 1930 for the first time. Dirac used this model to explain quantum states of negative energy in his equation and in order to justify relativistic electrons. Dirac

---

[10] - Chapter 15, pages 696-716, Bransden & Joachain, Quantum Mechanics, http://www.physics.udel.edu/~msafrono/PDF/L23.pdf

[11] - Carl David Anderson (1905 – 991), Discovery of the Positron, https://www.aps.org/programs/outreach/history/historicsites/anderson.cfm

[12] - Dirac P. A. M. "A Theory of Electrons and Protons" ROYAL SOCIETY PUBLISHING, 1930, doi: 10.1098/rspa.1930.0013

ratiocinated that all states of negative energy have been occupied by electrons in which they are not a part of the nature. It means that there exists a Sea of electrons with negative energy beyond the nature. He also ratiocinated that with a high energy photon, we can take apart an electron with negative energy from this Sea and convert it to an ordinary electron with positive energy.

Inexistence of negative energy means existence of positive energy, thus the hole behaves in a way that as if it is a particle with positive energy. On the other hand, inexistence of negative charge means existence of positive charge. This hole-particle alike electron has positive charge that was called positron.

In CPH theory, by defining the structure of photon, Dirac Sea is a physical fact that not only holds for positron but also it is inseparable part of the nature and even we can conclude Weyl fermions from it. Weyl fermions with spin $\frac{1}{2}$ have zero rest mass alike electrons (in quantum mechanics)[13]. In pair production of "electron-positron", was specified that the expression "negative energy" is not appropriate for these types of particles that later were called anti-particle. In fact, different electrical properties of electron and positron must be investigated in the structure of their producer that it means finding it in the structure of photon.

On the other hand, if a full energy photon (Gamma) that has this virtue that can be converted to two particles with different electric charges and all photons independent of their frequencies, carry electromagnetic energy. This virtue of electromagnetic energy must be investigated in electric and magnetic fields dependent to photon that can be converted to electron and positron with different electric charges.

2.4 From the Dirac equation to the photon structure

In pair production of "electron-positron", one photon with spin 1 and at least energy $E = 1.022\ MeV$ is converted to two fermions, electron and

---

[13] - Hamish Johnston, "Weyl fermions are spotted at long last"

http://physicsworld.com/cws/article/news/2015/jul/23/weyl-fermions-are-spotted-at-long-last positron with spin $\frac{1}{2}$, each of them with context of energy $0.511\ \text{MeV}$ in vicinity of a heavy nucleus so that we have the following relation:

$$\gamma \rightarrow e^- + e^+ \qquad (2\text{-}18)$$

Relation (2- 18) is justifiable according to Dirac equation by relations (2- 16 and 2- 17). In pair decay, an electron is combined with a positron and is produced two photons. In pair decay, reverse of relation (2- 18) takes place and we will have:

$$e^- + e^+ \rightarrow 2\gamma \qquad (2\text{-}19)$$

In all physical processes including pair production and decay, it must be held the following conservation laws:

1-   Electric charge conservation law, pure charge before and after the process must be equal.

2-   Linear momentum and total energy conservation laws: These rules has made forbidden production of just one photon (Gamma ray). Two photon with the same energy move but in two opposite directions (figure2-1). Angular momentum conservation law must be held too. In fact, in the process of "electron-positron" decay, these following relations hold:

$$e^- + e^+ \rightarrow 2\gamma$$

$$E_{2\gamma} = 2m_0 c^2 + E_{e^-} + E_{e^+}$$

$$m_0 c^2 = 0.511\ MeV$$

In which $m_0 c^2$ is zero rest mass of electron (also positron) and E_{e^-}, E_{e^+} are kinetic energy of electron and positron that are converted to energy of photons ($E_{2\gamma}$) at the time of pair decay.

Till here everything is true and it is justifiable and consistent with quantum mechanics laws (and also standard model). But an essential question is considerable. Before proposing the question, it is necessary to pay attention to physical phenomena by more accuracy and a different approach.

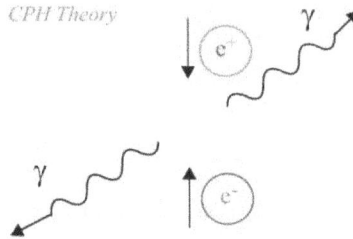

Fig2- 1; Pair electron – positron decay

In all these processes, matter is converted to energy and vice versa. Are conservation laws (as mentioned above) only related to matter and energy (photon) is just a part of this process or photon is one of the two main players of these processes? On the other hand, energy is converted to matter; matter has some properties including electric charge that energy apparently lacks it (because photon is electrically neutral) or these properties is transferred from energy to matter? In pair decay, these properties of matter including electric charge also is transferred by another method to the structure of energy (photon) or it is completely wiped out? In the sequel, it is tried to be answered to these questions by investigating some physical phenomena.

Quantum theory implies that a charged particle obtains energy while striking with a photon. In Compton Effect, photon with initial energy and momentum E_1, P_1 loses a part of its energy and moves with energy and momentum E_2, P_2 and this energy is transferred to charged particle[14]. Therefore, both particle and photon continue to their motions in paths that are not necessarily the previous paths (figure 2- 2).

It is a scientific fact that the vibrating motion of the atoms causes the cloud of electrons to oscillate and this oscillation generates electromagnetic radiation. Since all electromagnetic radiation travels at the same velocity

---

[14] - Derivation of Compton's Equation,
https://physics.ucsd.edu/students/courses/spring2015/physics4e/compton.pdf

the frequency and wavelength of the generated radiation depends on the frequency of the oscillating electron cloud[15].

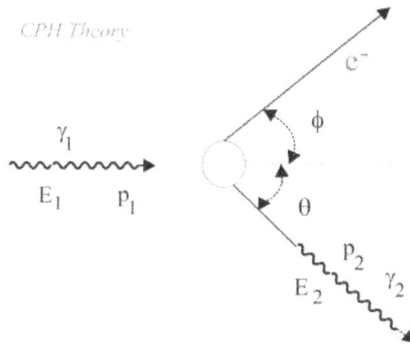

Fig2- 2; Compton Effect

However, if energy of photon increases in the path of its motion, like blue-shift phenomenon, the frequency of photon increases, therefore, photon is not a solid particle, it is formed of sub quantum energies, and interaction between sub quantum energies inside the photon is the main factor of its frequency. In other words, frequency of the photon is a function of interaction between its internal components of its structure. Therefore, the next step is to recognize sub quantum energies and their properties, so that it can be consistent with experimental conditions.

2.5 Dirac equation and Sub Quantum energy (SQE)

To explain and define sub quantum energy, it is necessary to analyze the relations (2- 15 and 2- 16). By taking square root from both sides of relation (2- 15), we will have:

---

[15] - Generation of Electromagnetic Waves.

http://xtide.ldeo.columbia.edu/mpa/Clim-Wat/Climate/lectures/energy/generation.html

$$E^2 = (mc^2)^2 \rightarrow E = \pm mc^2 \qquad (2\text{-}20)$$

In general state, equation (2- 20) does not accept any limitation for mass and energy regarding its value. Moreover, in limit of zero mass (zero rest mass of particles), Dirac equation was reduced to Weyl equation[16]. Weyl equation predicted the existence of fermions that their rest mass is zero[17], but they have spin $\frac{1}{2}$. Because here, the aim is to investigate and recognize the structure of photon. We reduce β matrix as follows and now we call it matrix A until after computations and necessary conclusions, we choose a special notion for it:

$$A = \begin{bmatrix} 1 & 0 \\ 0 & -1 \end{bmatrix} \qquad (2\text{-}21)$$

Therefore, the relation (2- 16) changes and converts as follows:

$$Amc^2 \rightarrow \begin{bmatrix} mc^2 & 0 \\ 0 & -mc^2 \end{bmatrix} \qquad (2\text{-}22)$$

According to relations (2- 16 and 2- 17) and in a special case that a photon collides with a heavy nucleus with at least energy $E = 1.022\ MeV$, we can write:

$$E_+ = mc^2, \quad E_- = -mc^2$$

That is called the process of pair production of electron and positron. Therefore, in general case, the relation (2- 22) is reagent of energy for two fermions with spin $\frac{1}{2}$ that one of the possible case describes pair production of electron-positron.

But occurring other cases is possible including photon with energy less than $E = 1.022\ MeV$ is decayed to two fermions with spin $\frac{1}{2}$, that move

[16] - William O. Straub, "WEYL SPINORS AND DIRACíS ELECTRON EQUATION" 2005, http://www.weylmann.com/weyldirac.pdf

[17] - Hermann Weyl, "GRAVITATION AND THE ELECTRON", PALMUR PHYSICAL LABORATORY, PRINCETON UNIVSRSITY, Communicated March 7, 1929

with speed of light in which it is describer of Weyl fermions[18] and they are called massless fermions or Weyl fermions[19] (or particles with zero rest mass).

According to Campton Effect and gravitational blue-shift, energy of a photon can decrease or increase without changing in its physical properties (except its energy and frequency). It means that whatever is increased to the energy of photon, it has the same total properties of photon (properties of electromagnetic energy). In other words, all photons have common physical properties except the value of energy that again it can be used the relation (2- 22) for them. Therefore, at least electromagnetic energy can be defined as follows:

$$E_{minimum} = \frac{hc}{\lambda_{max}} \text{ , where } E_{minimum} \text{ is detectable} \qquad (2\text{-}23)$$

According to relation (2- 22), $E_{minimum}$ includes two parts that it can be written as follows:

$$AE_{minimum} \rightarrow \begin{bmatrix} +\frac{E_{minimum}}{2} & 0 \\ 0 & -\frac{E_{minimum}}{2} \end{bmatrix} \qquad (2\text{-}24)$$

In relation (2- 24), the minus sign does not imply being negative of energy (or negative mass), as positron is not negative energy or mass in pair production. Signs $+, -$ in relation (2- 24) show electromagnetic fields around a charged particle and carry the same type of electromagnetic energy that there exists around a charged particle.

Therefore, the photon is formed of two types of positive and negative sub quantum energies that we show them by operators, right wedge ▷ for

[18] - Su-Yang Xu, et, at. "Discovery of a Weyl fermion state with Fermi arcs in niobium arsenide", Nature Physics 11, 748–754 (2015) doi:10.1038/nphys3437

[19] - Steven R. Elliott, "Colloquium: Majorana Fermions in Nuclear, Particle and Solid-state Physics", 2014, arXiv:1403.4976v2

positive sub quantum energy and left wedge ◁ that are defined as follows (see section1):

Positive Sub Quantum Energy; $SQE^+: \rhd= +\frac{E_{minimum}}{2}$ (2- 25)

Negative Sub Quantum Energy; $SQE^-: \lhd= -\frac{E_{minimum}}{2}$ (2- 26)

It is obvious that spin of sub quantum energy (SQE) is equal to $\frac{1}{2}$. In general case, relation (2- 22) can be written by using the definition of positive and negative sub quantum energies $\rhd$, $\lhd$ in which k is a natural number and instead of A, we use γ that is sign or symbol of electromagnetic energy:

$$\gamma = \begin{bmatrix} 1 & 0 \\ 0 & -1 \end{bmatrix} \qquad (2\text{-}27)$$

$$\gamma mc^2 \rightarrow \begin{bmatrix} k \rhd & 0 \\ 0 & k \lhd \end{bmatrix} \qquad (2\text{-}28)$$

In relation (2- 28), $k \rhd$ is positive virtual photon γ^+, in which carries positive electrical force and forms positive electric field and $k \lhd$ is negative virtual photon γ^- that carries negative electric force and forms negative electric field. Every real photon is formed of two virtual photons. Therefore, we will have:

$$\gamma^+ = k \rhd, \ \gamma^- = k \lhd \rightarrow \gamma = \gamma^+ + \gamma^- \qquad (2\text{-}29)$$

As charged particles absorb or repulse each other and are ineffective on neutral particles, homonymous virtual photons repulse each other, non-homonymous virtual photons absorb each other and they form quantum energies and it causes two non-homonymous charged particles accelerate towards each other. Virtual photons play role of force, force cannot be revealed. But you can see its effect. According to equation $F = -dU/dx$, only when you can speak about a force that you be able to detect the energy which is related to it.

2.6 Pair production and decay

The structure of photon and new approach on Dirac equation help us to understand and describe particle – antiparticle phenomena dipper than before.

In fact, this is useful key to open new windows on hidden parts of physics problems. For example, in pair production and decay of "electron-positron", a high energy photon $E = k(\rhd + \lhd)$ converts to e^+ and e^- (see figure 2- 3) that is given by;

$$E = k(\rhd + \lhd) \rightarrow e^+ + e^- = (e^+ = k\ \rhd) + (e^- = k\ \lhd) = k(\rhd + \lhd) \quad (2\text{-}30)$$

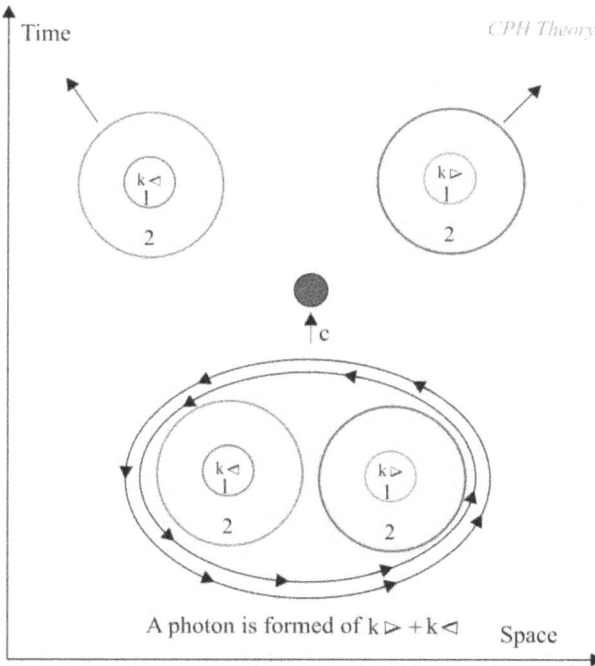

Fig2-3: In pair production, a high energy photon separates to electron-positron

For pair electron – positron decay to two photons (figure 2- 4) we have:

$$e^+ + e^- \rightarrow 2\gamma = \left(\frac{k}{2} \rhd + \frac{k}{2}\ \lhd\right) + \left(\frac{k}{2} \rhd + \frac{k}{2}\ \lhd\right) = k(\rhd + \lhd) \quad (2\text{-}31)$$

Attention to photon structure changes our perspective of charged particles. It also provides us with a new tool to be able to overcome physics problems in a better way. This approach will show us how particles are formed.

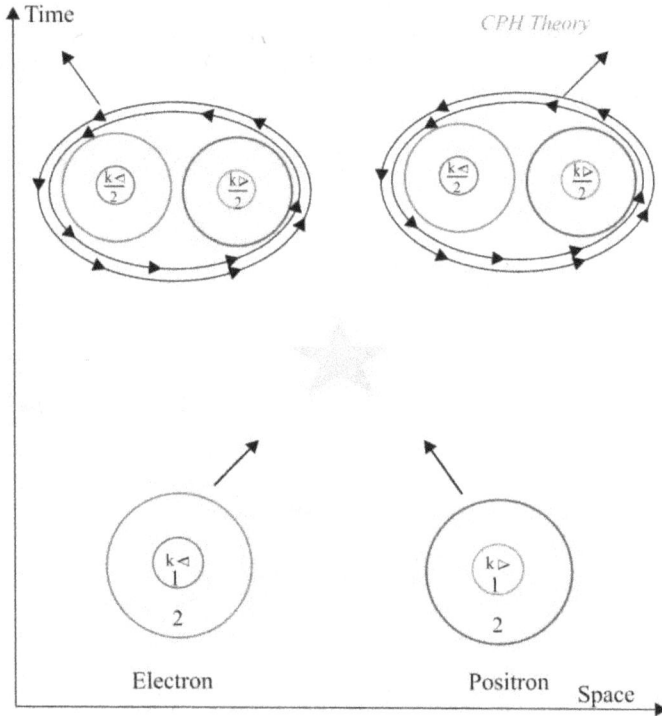

Fig2- 4; pair decay to two photons

Note that in the process of pair production of electron – positron, firstly the mass of photon is at least equal to the total mass of the electron and positron. Secondly, in the collision with nucleus, the photon is shattered due to caused impact of this collision. For three photons we can write;

$$e^+ + e^- \rightarrow 3\gamma = 3\left(\frac{k}{3}\triangleright + \frac{k}{3}\triangleleft\right) = k(\triangleright + \triangleleft) \qquad (2\text{- }32)$$

Now suppose that a photon with the energy less than required energy of pair production of electron and positron enters into the process of collapsing, consider the following process in equation (2- 30) and put the number k_1 instead of the number k, so that $k_1 < k$, in this unknown process we have:

$$E = 2k_1 E_{SQE} = k_1 \triangleright + k_1 \triangleleft \rightarrow fw^- + fw^+ \qquad (2\text{-}33)$$

Where fw^- , fw^+ can move at the speed of light and with zero rest mass (regard to definition in modern physics), these particles can be Weyl fermions.

On the other hand, the photon has no electric charge, but it is formed of electric and magnetic fields. These properties are acceptable when photon has been contained from two types of sub quantum energies and with various electromagnetic properties. According to this point of view in pair production and decay e^+, e^-, we can get interesting results. Before pair production, we have a photon and after pair production, we have two fermions (electron and positron) in which each of them have their own electric fields. It means that electron and positron produce virtual photons that can absorb each other. After pair decay, their electric fields are disappeared along with electron and positron. Therefore, it must be generalized the method of production and physical properties of fields from fermions to the structure of photon and vice versa. Also with such an approach, we can recognize the mechanism of electromagnetic interactions, use it to explain strong and weak interactions, and take a step towards to unify forces.

2.7 Sub quantum fields

According to above descriptions, electric and magnetic fields of particles are formed and moreover, photon is formed of two perpendicular electric and magnetic fields. Energy and momentum of photon is equal to summation of energy and momentum of its formed particles. If we indicate summation of transmission and non- transmission energy of particles with H , then we will have for graviton, color charge and magnetic color:

$$H_G = E_G = E_{GT} + E_{GS} = constant \qquad (2\text{-}34)$$

$$E_G = E_{G^+} = E_{G^-} = E_{G_m^+} = E_{G_m^-} = constant$$

For sub quantum energy:

$$H_{SQE} = E_{SQET} + E_{SQES} = constant \qquad (2\text{-}35)$$

For photon we can write:

$$H_{Photon} = 2nH_{SQE} = n(H_{SQE^+} + H_{SQE^-}) \qquad (2\text{-}36)$$

Also electric and magnetic energy of photon is obtained as follows:

$$E_{electric} = n(\kappa\, G^+ + \kappa\, G^-), \quad E_{magneticc} = n(G_m^+ + G_m^-) \qquad (2\text{-}37)$$

$$H_{photon} = 2n(\kappa + 1)E_G = 2n(\kappa + 1)H_G \qquad (2\text{-}38)$$

2.8 Sub quantum energy and Dirac Sea

If in the Dirac Sea, we use electric charge (in fact it is color-charge) instead of negative energy, then Dirac Sea is extensible to all physical phenomena including quantum vacuum, structure of quantum particles, materials and even stars and galaxies. Because, energy is formed of color-charges and magnetic-colors and by dense of color-charges and magnetic-colors, charged particles and neutral particles are produced. By combining relations (1- 41 and 1- 42), we have:

$$\nabla \times E + \nabla \times B \Leftrightarrow i(\boldsymbol{G^+}, \boldsymbol{G^-}) + j(G_m^+, G_m^-) \qquad (2\text{-}39)$$

For Zero point energy in vacuum, we can write:

$$(n\kappa G^+, n\kappa G^-) + (nG_m^+, nG_m^-) \Leftrightarrow (n\kappa G^+, nG_m^-) + (n\kappa G^+, nG_m^-) \Leftrightarrow (n \triangleright, n \triangleleft)$$

$$\Leftrightarrow (n \triangleright, n \triangleleft) \Leftrightarrow (\gamma^+, \gamma^-) \Leftrightarrow \gamma \qquad (2\text{-}40)$$

It shows that combining electric and magnetic field (even in vacuum) makes real photon.

2.9 Sub Quantum electrodynamics

Consider a charged particle (e.g. an electron) that creates an electric field around itself and constantly is spreading (propagating) virtual photons. The domain of propagation of this electric field is infinity. According to well-known physical laws, there is no change in the electrical charge and mass of charged particle by emitting virtual photons that carries electric force (and it carries electrical energy too). Therefore, we have a permanent machine in which we know its production, but we do not know about its mechanism and consumable and there is no information in this case. Just it is said that there is an electric field around any charged particle. How is created this field, what is its interaction with other electric and non-electric fields, including gravity, nothing is said, namely, there is no explanation.

Here according to the sub quantum energies $\triangleright$ and $\triangleleft$, the mechanism for generating electric fields, the dynamics of attraction and repulsion between charged particles are analyzed.

Electron is a set of negative color-charges that are preserved by electromagnetic field due to its surrounding magnetic-colors. This rotational sphere (spinning electron) is adrift (floating) in a sea of gravitons and as it already was explained, gravitons are converted to positive and negative color charges in vicinity of electron. There is same explanation for positron. Electron effects on existing color-charges around itself by having two special properties. Electron has continuous spinning state that can create an electric field that is formed of moving color-charges, then magnetic-colors are produced and then conditions are prepared to produce sub quantum energies. Positive color-charges are absorbed towards electron, but magnetic field around it is repellent of positive color-charges. By spinning movement of electron, a number of positive color charges are compacted and converted to positive virtual photon γ^+ and are repelled by its surrounding magnetic field. As the same way, positron absorbs negative color-charges and its surrounding magnetic field compacts negative color-charges and propagates it as negative virtual photon γ^-. Therefore, we can define an operator that expresses the process of producing positive virtual photons by electron. If we show this operator with $\lhd s$ that effects on electron and it is respect to time of γ^+ , it means that it creates the carrier of positive electromagnetic force, then we have:

$$\frac{d}{dt} \lhd s(G^+) = a \rhd = \gamma^+ \qquad (2\text{-}41)$$

Where a is a natural number. As the same way, positron behaves like electron that is similar to a generator and it produces and propagates negative virtual photons (Figure 2- 5) and then we have:

$$\frac{d}{dt} \rhd s(G^-) = a \lhd = \gamma^- \qquad (2\text{-}42)$$

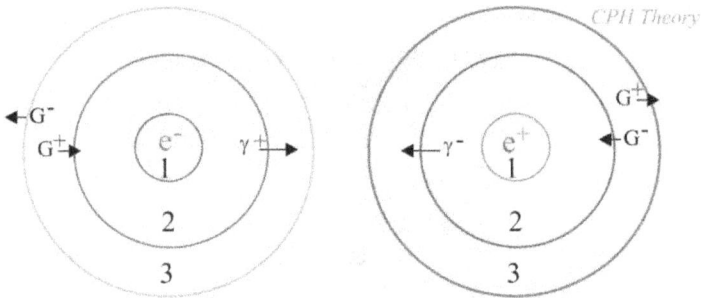

Fig2- 5: Electron and positron are attracted each other by positive and negative virtual photons.

When $a \rhd = \gamma^+$ from the electron reaches to area 2 of positron, it combines with $a \lhd = \gamma^-$, and a real photon is created. According to Newton's second law, what happens when (at the moment) γ^+ reaches the positron?

1-Two different virtual photons γ^+, γ^- absorb each other and the effect of magnetic field gradually neutralizes on γ^-.

2- Positron and γ^+ repel each other by the force F_{1e}

3- Positron and γ^- absorb each other by the force F_{2e} (because by reaching γ^+ to the area 2 around positron, the effect of propulsion (buoyancy) for magnetic field of positron neutralizes and cannot repulse γ^-. According to the figure (2- 6) and Newton's second law, we have:

$$F_{1e} = \frac{E(\gamma^+)e^+}{r_1^2}, \qquad F_{2e} = \frac{E(\gamma^-)e^+}{r_2^2}$$

$$r_1 > r_2 \Rightarrow F_{2e} > F_{1e}$$

$$F_{e=}F_{2e} - F_{1e} = m_e + a$$

Positron gets energy $\gamma^- + \gamma^+ = \gamma$ and accelerates by force $F_e = F_{2e} - F_{1e}$ toward the electron.

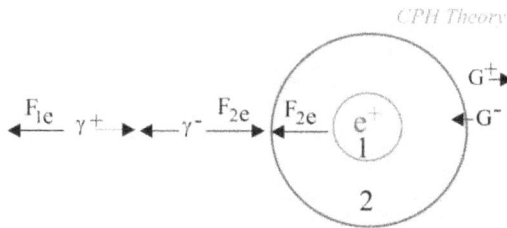

Fig2-6: Acceleration the electron and positron according to Newton's second law

The similar mechanism happens for electron, when $a \triangleleft = \gamma^-$ reaches from positron to area 2 around electron, combines with $a \triangleright = \gamma^+$ and creates a real photon. (Figure 2- 7). Then, we have:

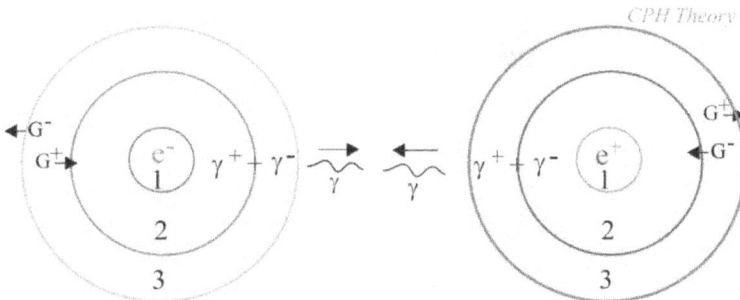

Fig2-7: Production and combination of virtual photons by electron and positron

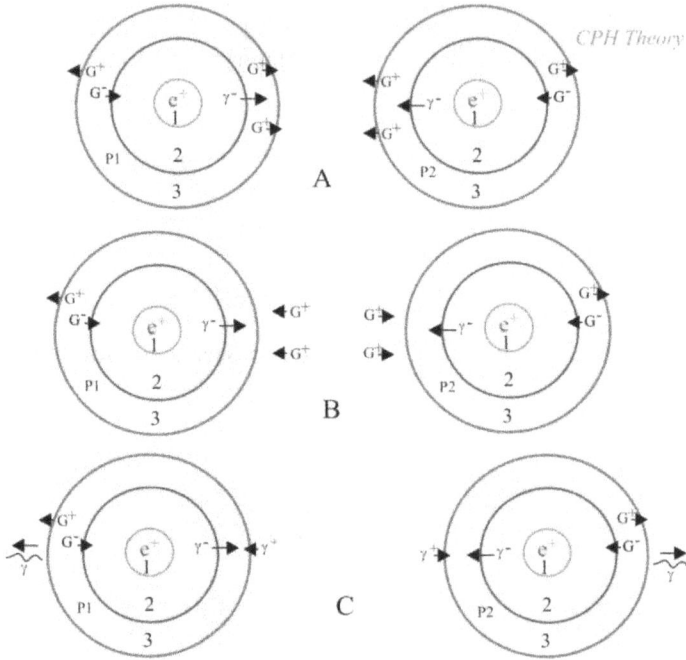

Fig2-8; two positrons repel each other by same color charges

For two same charged particles such as positrons, they repel each other by same positive color charges, their production's powers are same (figure 2- 8-A). When a lot of G^+s from P2 (positron 2) reach to near area 2 of P1 around γ^- (figure 2- 8-B), they form a positive virtual photon γ^+ (figure 2- 8-C). Then the γ^+ and γ^- combine with each other and a real photon appears. P1 absorbs energy and accelerates away from P2.

The same process happens for P2. Honestly, this process needs more research for better description and develop this idea. However, any acceptable mechanism should be able to describe the transferring energy and momentum for two same charged particles.

Note that electric force is almost $F_{electrcity} \cong 10^{40} F_{gravity}$ gravitational force[20], so the effect of color-charges on the virtual photon is not significant.

Electron creates positive electric field, positron creates negative electric field and they propagate these fields in the space. When charged particle was discovered, it imagined that the charged particle and the surrounding fields are the same. If we accept, the charged particle and the surrounding fields are not the same; it does not create any effect on electromagnetic theory, because two homonymous charged particles repel each other and two non-homonymous charged particles absorb each other. Then our observations shows that existing field around electron repels negative charged particle. Therefore, positive virtual photons are created around electron and negative virtual photons are created around positron. Now consider a positron that lies in electric field of an electron. Positron produces γ^- and at that moment γ^+ reaches to a positron. In this case, γ^- and γ^+ combine with each other and a quantum electromagnetic energy produces and is absorbed by positron. The direction of energy that is absorbed by positron is exactly towards electron and positron accelerates towards electron. A similar process happens for electron and this process is continuously iterated (figure 2- 5). Here it was considered just a path, it was assumed that the positive virtual photon moves on a specified path and goes from the side of electron toward positron and combines with negative virtual photon produced by positron and accelerates to positron that is not apparently consistent with quantum mechanics. Because in classical mechanics, just a path indicates the motion of the particle, while all paths for a particle in quantum mechanics can be considered, even routes that is similar to the classic route. However, it is not true, a positive virtual photon can move on all possible routes to reach positron or not. With the amazing advent of quantum mechanics, still classic patterns are used in the design of quantum computers[21]. Because our mind can well analyze classical

---

[20] - Michael's Question, "Electric Forces Versus Gravitational Forces"
http://www.batesville.k12.in.us/physics/phynet/e%26m/electrostatics/michaels_question.htm

[21] - Classical Mechanics Helps Control Quantum Computers, I-Connect007, July 6, 2017 | Technical University of Munich

phenomena, but the classic and quantum phenomena follow the same rules. In addition, the classical and quantum classification of physical phenomena is a human approach. It is important that not only electron is producing and emitting positive virtual photons continuously, but also a lot of positive virtual photons are moving in electrical field of electron, each of them has been entering to area 2 of positron, it would do the same action as described above. It is important that we understand the mechanism of this action and explain in a way that is consistent with the basic laws of physics.

For example, look at forming snowflake. Due to perching in different conditions while forming, snow crystals possess complicated shapes with many details and then due to existence of many probable states, probability of finding two snow crystals with completely similar structure is very poor. The researches of Kenneth Libbrecht[22] on snow crystals in different temperatures shows complexity of its geometric shapes. These different conditions can be categorized in this way. The different number of water molecules in crystals, the difference of gravitational force imposed on the snowflakes and the forces of snowflakes around one snowflake impose on it before freezing effects on geometrical shapes of snow crystals.

As the same way, we do not know and we cannot observe or compute that which positive virtual photon goes forwards positron, but we know from physical laws that each positive virtual photon that approaches (comes close to) the positron, positron accelerates towards electron. Understanding and explanation of this matter is very important for us, not large number of virtual photons and different paths probably go through. Even in Feynman diagrams, it is important that what the results of interaction between particles is, not probability of traveled and passed paths. According to $\gamma^+ + \gamma^- = \gamma$, it can be explained very well that why carrier photons of electromagnetic force are not visible. The production process of virtual photons by using sub quantum energies is explainable. This explanation is based on physical fact that a photon is formed of two sets of positive and negative color-charges in which each of them has its own dependent

http://ein.iconnect007.com/index.php/article/105177/classical-mechanics-helps-control-quantum-computers/105180/?skin=ein

[22] - Kenneth G. Libbrecht,, professor of physics at the California Institute of Technology "Snow Crystals" http://www.snowcrystals.com

magnetic field. Production of virtual photons occurs like radiation of charged particles with a specific frequency. It means that any charged particle with just one turn of a spinning movement creates and emits a virtual photon. In fact, every turn in spinning movement of charged particle can be considered like oscillations that causes radiation and creates a virtual photon. Virtual photons move with linear speed c and form electric field around a charged particle. Any charged particle that lies in this field, will have interaction with existing virtual photons in this field.

Note: According to above descriptions, it is observed that the energy is generated by field, and matter is generated by energy, so we can say that in CPH theory, energy is intensive field and matter is a dense energy.

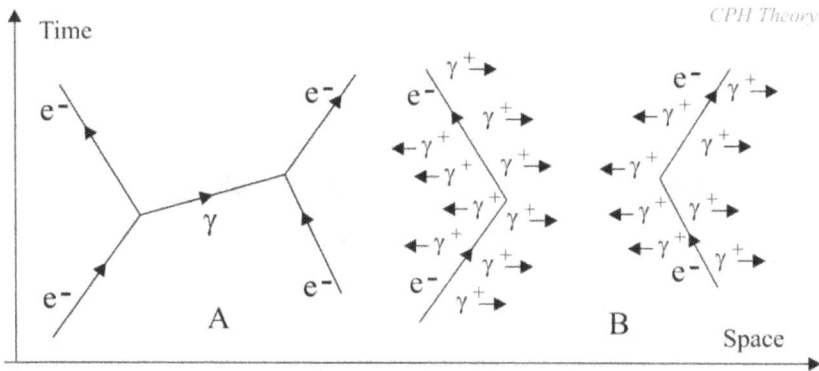

Fig2- 9: Feynman diagram A and virtual photons B for repulsion of two electrons

2.10 Sub quantum energy and Feynman diagrams

In quantum electrodynamics, charged particles (for example electron and positron) have interaction with each other through propagation and absorption of photon (particles that carry electromagnetic force) and these interactions are justified by Uncertainty Principle. Even Feynman diagrams is a representation to describe physical processes. While by using sub quantum energies and positive and negative virtual photons, interaction between charged particles is explainable as physical analysis and mathematical computations. For example, notice to repulsion of two electrons (figure 2- 9) and absorption of positron and electron (figure 2-10).

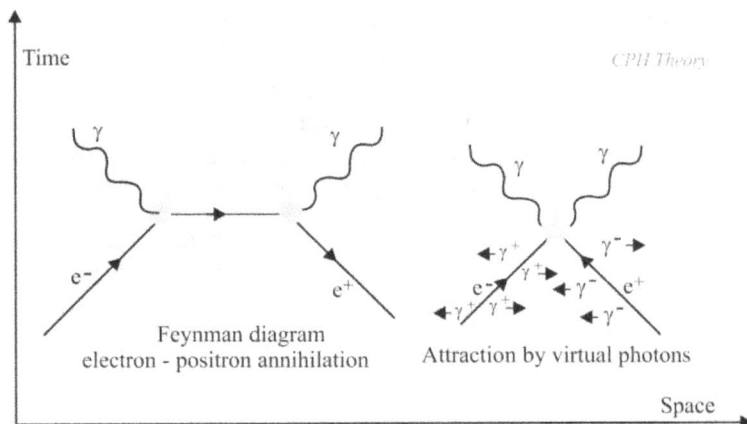

Fig2- 10: Feynman Diagrams and absorption of electron and positron by virtual photons

2.11 Absorption and emission photons by electron

What we know from quantum mechanics is: When an electron absorbs energy, it jumps to a higher orbital. An electron in an excited state can release energy and 'fall' to a lower state[23]. But experiment shows absorption is not so easy and photon has shape and photons are some 4 meters long which is incompatible with unstructured concept[24]. And probability that a single Rubidium atom would absorb a single photon of either type was just over 4%[25].

According to the quantum mechanics that photon and electron are unstructured particle, we cannot describe the mechanism of absorption and emission photon by electron that is logically unacceptable. As it was

---

[23] - The Bohr Atom,
https://chem.libretexts.org/Core/Physical_and_Theoretical_Chemistry/Quantu
m_Mechanics/10%3A_Multi-electron_Atoms/The_Bohr_Atom

[24] - Victor Leong, et, at. Time-resolved Scattering of a Single Photon by a Single Atom, 2016, https://arxiv.org/pdf/1604.08020v1.pdf

[25] - Last reference

described photons are combination of positive and negative virtual photons. Photon is a very weak electric dipole that is consistent with the experience and this book is asserted.

When photon reaches to area 2 of electron, positive side of photon does change the shape of electron, electron absorbs it. But electron is moving and in a fraction of a nanosecond emits photon, but not the necessarily exactly opposite direction of absorption, because in during the absorption and emission, electron has shifted. It is considerable that for a beam of light (a lot of photons), movement of electrons are not detectable (figure 2- 11).

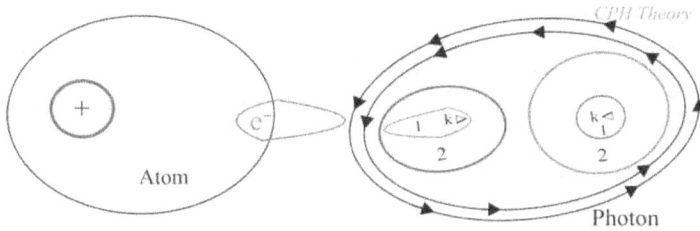

Fig2- 11: absorption photon by electron

This property of photon (very weak electric dipole) can describe the absorption and emission energy by charged particles.

Emission; Take a look at the figure (2- 11), for a high energy photon, electron leaves the atom (such as the photoelectric effect). But for a low-energy photon, photon is became a part of energy of electron. Electron is not a spherical rigid and is moving around nucleus, distance between electron and nucleus is changeable. Electron is unable to keep photon and radiates.

3

From sub QED to sub QCD and Higgs boson

Since the 19th century, some physicists have attempted to develop a single theoretical framework, which can be applicable for the fundamental forces of the nature a unified field theory. The classical unified field theories attempt to create a unified field theory based on the classical physics. In particular, several physicists such as Faraday[1], Planck[2] and Einstein actively were pursued unification of the gravitation and electromagnetism. Einstein believed that there was a link between the need to resolve apparent paradoxes of the quantum mechanics and the need to unify electromagnetism and gravity[3]. The classical unified field theories were unsuccessful.

---

[1] - Michael Faraday, "On the possible relation of gravity to electricity" Royal Society 1850. Printed in vol. 141 of the Philosophical Transactions (1851), pp. 1-6

[2] - Brandenburg, J. E., "Pure EM (Electro-Magnetic) Propulsion and Theories of Gravity-EM Unification", 2010, Orlando, Florida, http://enu.kz/repository/2010/AIAA-2010-1610.pdf

[3] - Einstein's quest for a unified theory, This Month in Physics History, APS Physics; [Online] available; http://www.aps.org/publications/apsnews/200512/history.cfm

3.1 A Theory of Everything

A Grand Unified Theory (GUT) unifies the three forces described by the Standard Model of particle physics - the electromagnetic, weak, and strong force - into a single force that breaks into the other three at low energies. The electromagnetic and weak forces are already unified into one force, the electroweak force, so all that remains is to unify the third (strong) force with the other two. A Theory of Everything is literally a theory of everything, including the force of gravity, and anything else in our universe that our current theories cannot explain[4].

It may be thought that these two problems are separate of each other, but both problems; GUT and the Theory of Everything has a common root. Therefore, the solution of each of them includes solution to other problem as well.

A.   The solution of GUT: if we describe the mechanism of the virtual photons production (electromagnetic force carrier) by charged particles, then we will see that electromagnetic repulsive force in a very short distance, turns to the attractive force, then the GUT problem can be solved. In this way we will reach to unify the electromagnetism and gravity that is the Theory of Everything.

B.   The solution of the Theory of Everything: to get understand the Theory of Everything, we must re-define fundamental particles. In CPH Theory, mass/energy and the amount of speed of fundamental particle must be constant and not turn into other particles. While in the Standard Model, fundamental particles have variable mass and speed, so they are not fundamental particles. To find fundamental particle we must reconsider and analyzed the interaction between photon and the graviton. In this way the GUT problem can be solved, too.

Due to this reason, we need to generalize the Dirac's equation and Sea. The Dirac's Equation is usually limited to the high energy photons and the pair production and decay of a particle – antiparticle, while Dirac's Sea can

---

[4] - What are- he Grand Unified Theory and the Theory of Everything and what is the difference between them?

https://www.quora.com/What-are-the-Grand-Unified-Theory-and-the-Theory-of-Everything-and-what-is-the-difference-between-them be used for all quanta of energies. With the generalization the Dirac's Equation and Sea, the similarity between quantum chromodynamics (QCD) and photon can be resulted. In other words, there are color properties not only of in quarks, but the existences of their properties are acceptable in photon structure and even in a gravitational field, too.

In dealing with the interaction between charged particles (especially, two same charged particles) two cases are remarkable and reviews.

1- Production binding energy between charged particles (QED), especial in structure of nucleons (QCD).

2- Input and output particles in interaction charged particles process.

Two above cases, in two separate theories of quantum electrodynamics (QED) and quantum chromodynamics (QCD) that describes the strong interactions)[5] are investigating.

In generally, to combining QED and QCD, we cannot ignore gravity. In section 2 QED was described. In this section QCD is described that is showed how two same charged particles repel each other in great distance and absorb each other at a very small distance. First of all, it should be noted that there is no force in the classic sense in nature that has mentioned in last section. Due to this reason, a fundamental force is just binding energy between fermions such as quarks. This binding energy is electromagnetic energy that is called photon. In quantum mechanics, tiny packages of electromagnetic energy called photons and the force carrier for the electromagnetic force (even when in static case via virtual photons are applied). But there is difference between real photon (package wave of electromagnetic energy) and virtual photon that the force carrier for the electromagnetic force in CPH Theory.

3.2 Sub Quantum Chromodynamics

As we know in quantum mechanics, there is a strong interaction in nucleus of an atom and its range is short and less than the radius of an atom. Carrier

---

[5] - QUANTUM CHROMODYNAMICS, Revised 2015 by S. Bethke (Max-Planck-Institute of Physics, Munich), http://pdg.lbl.gov/2015/reviews/rpp2015-rev-qcd.pdf

of the strong interaction force that is called gluon is a particle with spin one (spin of photon is one, too).

Strong interaction is the factor of integration among large number of positive charged particles in nucleus of atom[6]. Therefore, it is stronger than electric repellent force among existing positive charged particles in the nucleus. For example, proton is formed of 3 quarks, two up quarks (u) with $(+\frac{2}{3})$ electric charge and a down quark (d) with $(-\frac{1}{3})$ electric charge that is shown with p(udu). The subject that a down quark absorbs by up quark is justifiable with electromagnetic theory, because they have opposite electric charge. However, the subject that how two quarks gather together with homonymous charged particles is another problem that still there is some theoretical problems[7] and intuitive justification about that in modern physics that can be consistent with experiments.

The explanation is given in modern physics is that boson (gluon) with spin one is carrier of color charge force between quarks and it is stronger than electric force. However, the reason and mechanism of strong interaction is easily explainable by using sub quantum energies.

In general state, suppose two same electric charged particles A^+, B^+, lie in a bigger distance of the radius of proton. As explained in the previous section, each positive charged particle repels positive color-charges and absorbs negative color –charges (figure 3- 1). The magnetic field around it compacts these negative color-charges and emits it as negative virtual photon in the space. When the distance between these two particles is high (greater than the radius of nucleus of atom), before that emitted negative photon γ^- reaches from second particle to first particle, repelled positive color-charges by first particle have left the environment (they have got away from the charge surroundings). While in short distances, the repelled

---

[6] - http://www.britannica.com/science/nucleon

[7] - Matthew Francis, "Glueballs are the missing frontier of the Standard Model", Ars Technica, 2015
http://arstechnica.com/science/2015/10/missing-glueballs-are-sticky-problem-for-particle-physics/

K.A.Ter-Martirosyan, "Gluons in the QCD bound state problem - a way to exact solution", 2000 http://arxiv.org/abs/hep-ph/0007361

positive color-charges by a particle combines with negative color-charges around another particle and create electromagnetic energy.

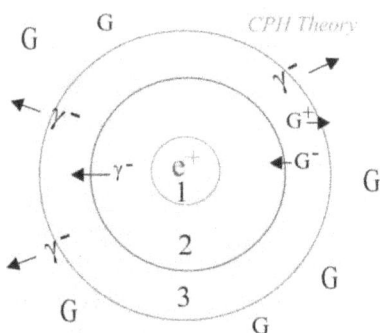

Fig3- 1; Locations around each positive charged particle

Suppose the particle A^+ produces a negative virtual photon γ^- in the time dt, it repels a number of positive color charges that can produce a positive virtual photon γ^+. If we consider the distance between these two particles, supposing speed of γ^- is at least equal to speed of light c, if $d > cdt$ (figure 3- 2), the repelled positive color-charges by each particle is ineffective on negative color-charges around the second particle. If $d < cdt$, the mechanism of attraction and repulsion of color-charges by each particle interfere with the mechanism of the other particle, positive and negative color-charges are converted to electromagnetic energy and these two particles absorb each other. Because if $d < cdt$, the binding energy between two particles A^+ and B^+ is stronger than repulsive electrical force between them. But if $d = cdt$, then the electrically charged particles are neural with respect to each other (figure 3- 2), which can produces vector bosons (weak nuclear interaction), so behavior of electromagnetic and weak nuclear interactions are very similar. This process can be explained as next part.

3.3 Gluon as a binding energy

Consider a positive charged particle that is adrift in a gravitational field. In fact, all charged particles are drift in a gravitational field. Gravitational field is made of infinite number of gravitons that are moving faster than speed of light c. Gravitons show color-charge properties in vicinity of charged particle or its magnetic field. Positive charged particle absorbs negative color-charges and repels positive color-charges. Therefore, in an area around positive charged particle is created a color-charge field and from a specific distance, positive color-charges cannot come closer to the charged particle. In area 3 (figure 3- 1), gravitational field is formed of gravitons. Gravitons are moving with linear speed $V_{GT} > c$. In area 3, they get color-charge properties. However, gravitons in any gravitational field can be converted to color-charge, but in vicinity of charged particles or electromagnetic fields are converted to color-charge with rapidly. Positive color-charges cannot enter into area 2 and are repelled, but negative color-charges enter into area 2, are compacted under the effect of magnetic field and are emitted as negative virtual photon γ^- and electric field creates.

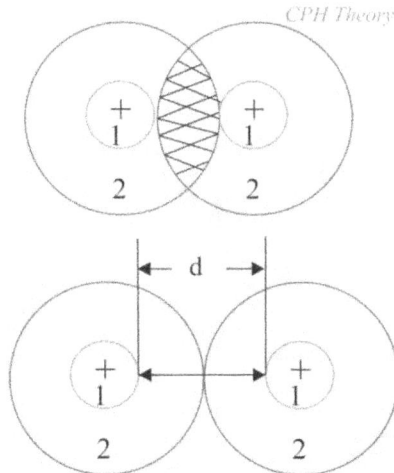

Fig3- 2; interconnect two positive charged particles

Now suppose two positive charged particles A^+ and B^+ are near each other that location2 interferes with each other (figure 3- 2). Therefore, in location2, positive color-charges G^+ from A^+ and negative color charges G^- from B^+, have the same direction movement that is toward the B particle. They combine and convert to electromagnetic energy and transfer to the particle B^+. The same action happens for positive color-charges G^+ from B^+ and negative color-charges G^- from A^+; so, they form quantum energy that moves toward A^+. So, the binding energy between A^+ and B^+ is produced.

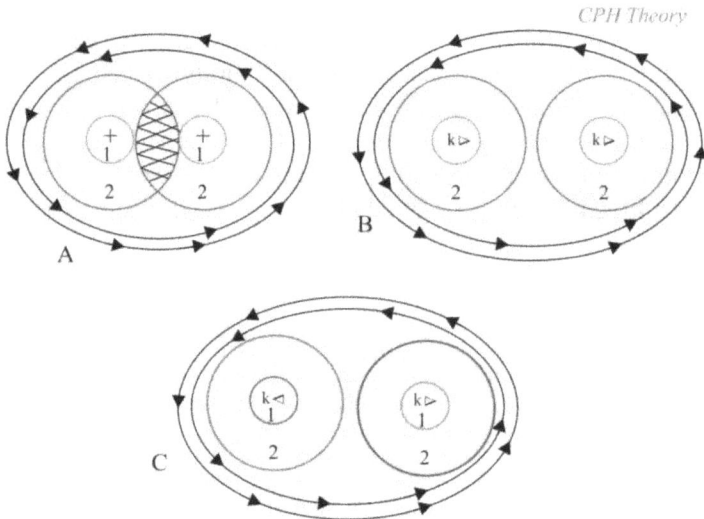

Fig3- 3: A. The magnetic field around two same charged particles.

B. Magnetic field around two same *SQEs*.

C. Photons are formed of ▷ $+k$ ◁ , magnetic fields around k ▷ and k ◁ prevent them from this combination

Nuclear fusion in the center of stars is repeating this process. When two homonymous charged particles became close enough to each other , their magnetic fields are united and keep togther these homonymous charged particles like plasma of charged particles (figure 3- 3- A). In the center of the stars, due to high speed (transitive energy) of nuclues of atoms, they come close enough together and protons (in fact quarks) fall in each other color-charges areas and provide the necessary binding energy and nucleuses do fusion. There are many protons (in fact quarks) in a heavy nucleus, the number of quarks can have common color-charges area and absorb each other.

Paying attention to internal structure of photon is very useful and important for better understanding of QCD and QED.

Mass-energy equivalence includes the concepts and applications beyond the concept of converting mass to energy and vice versa. Something that occurs from the interactions between quarks in the structure of protons is the logical result of the interaction between $\triangleright$ *and* $\triangleleft$ in the structure of photon. In addition, during conversion of energy into mass, properties of interactions between $\triangleright$ and $\triangleleft$ are transferred from structure of the photon to particles and anti-particles. The same process that happens for two non-homonymous charged particles (in the nucleus of atoms) in center of stars, happens for formation of the negative and positive virtual photon γ^-, γ^+ by negative and positive sub quantum energies $\triangleleft, \triangleright$ (figures 3- 3- B and 3- 3- C).

3.3.1 Weak nuclear interaction

For $d = cdt$ (figure 3- 2), the electrically charged particles are neutral with respect to each other, but the density of color - charges in this case is very high which vector bosons (weak nuclear interaction) is created, so behavior of electromagnetic and weak nuclear interactions are very similar. This process can be used to explain the weak interaction for following relations:

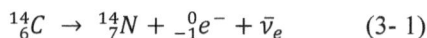

$$^{14}_{6}C \; \rightarrow \; ^{14}_{7}N + \, _{-1}^{0}e^- + \bar{v}_e \qquad (3\text{- }1)$$

Beta minus decay: A neutron decays into a proton, and an electron and an antineutrino are emitted.

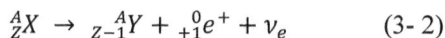

$$^{A}_{Z}X \; \rightarrow \; _{Z-1}^{A}Y + \, _{+1}^{0}e^+ + v_e \qquad (3\text{- }2)$$

Beta plus decay: a proton decays into a neutron, and a positron and a neutrino are emitted instead.

3.4 Higgs boson and CPH theory

The Standard Model of particles is unable to explain how particles actually get their mass. Physicists believe that the Higgs boson is a particle associated with the Higgs field, the mechanism through which elementary particles gain mass. According to standard model, without the Higgs field, atoms would not form. In 2012, scientists on two international experiments at the Large Hadron Collider at CERN laboratory announced the discovery of the Higgs boson[8]. While developing quantum chromo-dynamics, some new points of view of Higgs bosons have been introduced and also have been put into discussion about the other types and some specifications of Higgs bosons[9]. In the published articles in the recent years, most attractions have been noticed toward Higgs charges. Most of them have been paid attention to the Higgs bosons and electro-weak bosons but no anymore relations between gravity and Higgs has been said or noticed, yet[10]. In addition, observation of charged Higgs boson at (LHC) is beyond the Standard Model.

"A charged Higgs boson, $H^{\pm}$, is predicted in many models of new physics, with and without Supersymmetry (SUSY). The observation of a $H^{\pm}$ at the Large Hadron Collider (LHC) is thus expected to provide concrete evidence of physics beyond the Standard Model (SM)"[11].

---

[8] - Fermilab, Key Discoveries

http://www.fnal.gov/pub/science/particle-physics/key-discoveries.html#discover-higgs

[9] - V. Ravindran, Higher-Order threshold effects to inclusive processes in QCD, 2006

https://arxiv.org/pdf/hep-ph/0603041.pdf

[10] - Petra Häfliger; Michael Spira, Associated Higgs Boson Production with heavy Quarks in e+e- Collisions: SUSY--QCD Corrections, 2005 ,

https://archive.org/details/arxiv-hep-ph0501164

[11] - Rikard Enberga, et, at. Charged Higgs boson in the W± Higgs channel at the Large Hadron Collider, ScienceDirect, 2015,

http://www.sciencedirect.com/science/article/pii/S0550321315000383

3.4.1 What is the Higgs Mechanism?

For an example of spontaneous symmetry-breaking, imagine a complex scalar field whose value at any point in space is;

$$H(x, y, z) \qquad (3\text{-}3)$$

Consider giving the field a potential energy of the form;

$$V(x, y, z) = (|H(x, y, z|^2 - v^2)^2 \qquad (3\text{-}4)$$

Integrated over space. Here, $V(x, y, z)$ is a potential energy and $H(x, y, z)$ is Higgs field which is non-negative. There is a continuous manifold of minima at:

$$|H|^2 = v^2 \qquad (3\text{-}5)$$

What this means in less technical terms is that the potential energy density, as a function of H looks like the bottom of a bottle: a hump in the middle and a circular valley around it[12].

According to above relations, there is a Higgs particle in any small volume of space, which in this book is called Creative Particle Higgs or CPH.

3.4.2 Creative Particles of Higgs (CPH)

Definition: Suppose that there is a particle (smaller than a photon which can be located inside the photon) with constant mass m is moving with speed of V_{CPH} in any inertial reference frame and $V_{CPH} > c$ (c is the speed of light). So, the linear momentum of a CPH can be written as V_{CPH}, a CPH can be shown with zero Higgs boson H.

When a CPH has spin, it is called a graviton. Now as space is full of gravitons, so it can be said that space is full of CPHs.

---

[12] - Vincent S Ryan, Higgs mechanism, SCRIBD,
https://www.scribd.com/document/514522/Higgs-Bosons

3.4.3 Principle of CPH

A CPH is a particle with constant mass m which moves with a constant magnitude of speed, which equals to V_{CPH}. The CPH has a momentum of inertia **I**. In any interaction between the CPH and the other existing particles, the magnitude of V_{CPH} is constant and it does not change. Therefore;

$$\nabla V_{CPH} = 0 \ , \ in \ all \ inertial \ reference \ frames \ in \ any \ space \quad (3\text{-}6)$$

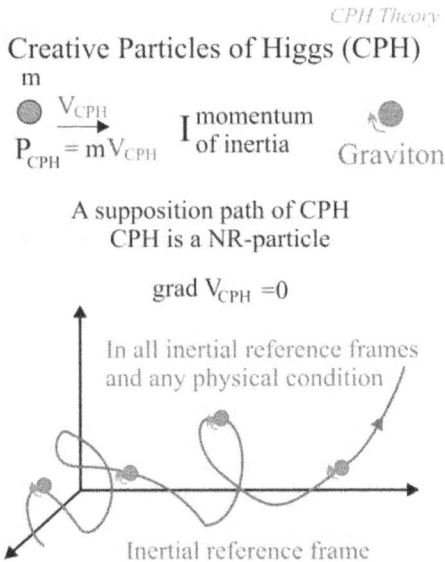

CPH Theory

Creative Particles of Higgs (CPH)

m

V_{CPH}

$P_{CPH} = mV_{CPH}$ I momentum of inertia   Graviton

A supposition path of CPH
CPH is a NR-particle

grad V_{CPH} =0

In all inertial reference frames
and any physical condition

Inertial reference frame

Fig3- 4; CPH has two movements; transfer motion and spin motion

As it has mentioned above, when a CPH has spin, it is called a graviton. So above principle can be comparable with relations (1- 27, 1- 28 and 1- 29) as follow:

$$|V_G| = |V_{GT}| + |V_{GS}| > |c| \quad (1\text{-}27)$$

$$\nabla V_G = 0 \ , \text{in all inertial reference frame and any space} \quad (1\text{-}28),$$

$$E_G = E_{GT} + E_{GS} = constant \qquad (1\text{-}29)$$

In gravitational blue-shift, energy of photon increases. Changing the photon energy in a gravitational field that is associated with the intensity of electric and magnetic fields of photon. It means that gravity works on the photon and gravitons enter to the structure of the photon in which it is justified with $F = -dU/dt$ equation. Also, not only a photon at lowest level of its energy is formed by some of the gravitons, but also its formed members have electric and magnetic properties that is called color-charge and magnetic-color in CPH theory. See figure (3- 5):

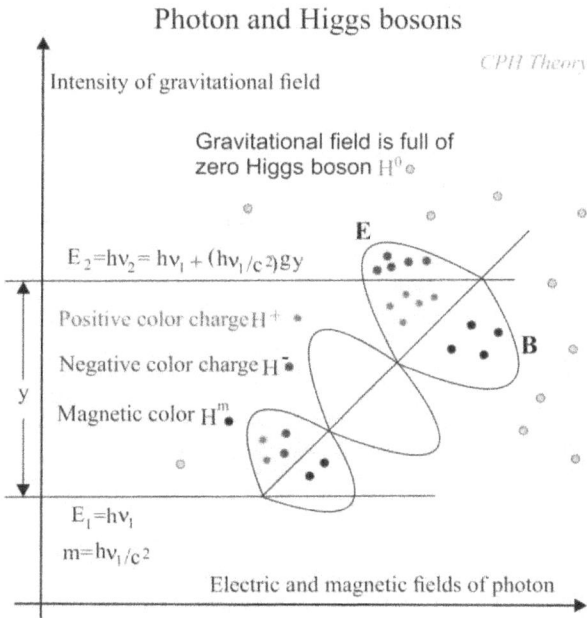

Fig3- 5; increasing intensity of electrical field and magnetic field during photon fall in gravitational field due to increasing of photon energy

According to CPH matrix (relation 1- 24) and figure (3- 5), then we can show:

$$CPH = \begin{bmatrix} \kappa\, G^+ & \kappa\, G^- \\ G_m^+ & G_m^- \end{bmatrix} \qquad (1\text{-}24)$$

$H^0; G, where\ G\ is\ symbol\ of\ graviton$

$H^+; G^+\ where\ G^+\ is\ positive\ color\ charge$

$H^-; G^-\ where\ G^-\ is\ negative\ color\ charge$

$H^m; G_m^+\ and\ G_m^-\ are\ magnetic\ colors$

Therefore, photons gain mass of gravitons (color charges and magnetic colors), see relation (1- 54). And photons (electromagnetic energy) convert to matter – antimatter that is given by;

Gravitations $\Leftrightarrow photons \Leftrightarrow$ matter $-$ antimatter   (3-7)

Also the gravitons have three motions inside of photon; one is self-spin, the other is spin from photon spin and the last one is linear speed equal to c. The theory of matter and its interactions is Standard Model cannot be the complete description of the universe. For example the mass of all fundamental particles must be less than the electroweak mass, whose value of a few hundred GeV/c2 follows from the physics of the Higgs boson[13]. The ATLAS and CMS experiments at CERN's Large Hadron Collider announced they had each observed a new particle in the mass region around 126 GeV. This particle is consistent with the Higgs boson predicted by the Standard Model[14].

Although the Standard Model describes the phenomena within its domain accurately, it is still incomplete. Perhaps it is only a part of a bigger picture of the modern physics which includes the deeper and hidden layer of subatomic world that has been dipped into the darkness of the universe[15].

---

[13] - David Curtin, Hidden worlds of fundamental particles, Physics Today, June, 2017
http://physicstoday.scitation.org/doi/10.1063/PT.3.3594

[14] - The Higgs boson, CERN Document Server, https://home.cern/topics/higgs-boson

[15] - The Standard Model, CERN Document Server,
http://home.cern/about/physics/standard-model

The question is, where is the hidden part of modern physics? Hidden part of modern physics lies beyond the uncertainty principle. Included in the sub quantum scale, where quantum interactions between photons and gravitons done. Hidden and dark side of modern physics is also a place where charged particles absorb and emit energy quanta, without any description of the mechanism of absorption and emission by charged particles. In modern physics, a charged particle creates an electric field itself, but the mechanism of this process is ambiguous and does not explain how a charged particle creates an electric field?

And above all, what is the scientific and precise definition of a fundamental particle? If we want to reach the different results, we need to change our thoughts and scientific beliefs.

So this question arises: what is the properties of a particle that could be really a fundamental particle? In CPH theory, a fundamental particle is a particle that is not decayed under any condition or is not convertible into other particles. Such a particle must be constant mass (energy), therefore, the value of speed must not change.

By this definition of fundamental particles, that standard model presents, particles are not fundamental, because their masses are not constant and they are convertible to energy. For instance, electron and positron absorb each other and convert to energy. This phenomenon holds for other fundamental particles in standard model even for photon, because energy of photon is variable (for example in gravitational field and Compton Effect) and in pair production, a high-energy photon converts to electron-positron. As the same way, it can be shown that even photon experiences time passing. In fact, a fundamental particle must not experiences time passing, and all other particles are made of it even quantum fields (see section 6).

This is reasonable that photon and graviton are not massless particles. They are NR-particles that have defined in section one.

3.5 Matter-antimatter problem

In standard model of particle physics, any particle has its ant-particle (In fact, matter and antimatter are identical). However, standard model cannot answer to this question why in our universe the amount of matter is more than anti matter. In other words, around us, there exist lots of matter, but antimatter does not observe[16]. However, in CPH theory, as has described in last section, the symmetry between positive and negative color-charges is important, not the number of equality between the number of particles and anti-particles.

The subject of existence of matter - antimatter was raised by Dirac's Equation and Sea and later it was confirmed experimentally. But what has been ignored here, is that photon/energy converts to particle - antiparticle and this is the property of energy that can be converted into matter and antimatter, and the reason of this property can be found in the structure of photon. In CPH Theory, any real photon is made up of two positive and negative virtual photons. Even in generation of a pair electron - positron, positive virtual photon converts to positron (according to negative energy in the Dirac's Sea), and negative virtual photon converts to electron. Any positive virtual photon is formed of a number of positive sub quantum energies, and any negative virtual photon is formed of a number of negative sub quantum energies, too.

According to equations (1- 25 and 1- 26), the number of positive color-charges and magnetic-colors in both positive and negative sub quantum energies ($\triangleright$ and $\triangleleft$) are equal. With the generalization of the Dirac's Equation and Sea, in any physical process equality the number of positive and negative sub quantum energies is provable, and there is no need to equality of matter and antimatter. Just it is important that the law of conservation of color – charge not to be violated that it completely holds.

Example1: Pair production and decay:

$$E = k(\triangleright +\triangleleft) \; \rightarrow \; e^+ + e^- = (e^+ = k \; \triangleright) + (e^- = k \; \triangleleft)$$

$$e^+ + e^- \rightarrow 2\gamma = \left(\frac{k}{2} \triangleright +\frac{k}{2} \triangleleft\right) + \left(\frac{k}{2} \triangleright +\frac{k}{2} \triangleleft\right) = k(\triangleright +\triangleleft)$$

---

[16] - Natalie Wolchover, 2016, http://www.livescience.com/34052-unsolved-mysteries-physics.html

$$e^+ + e^- \rightarrow 3\gamma = 3\left(\frac{k}{3} \triangleright + \frac{k}{3} \triangleleft\right) = k(\triangleright + \triangleleft)$$

Example 2: Proton and antiproton:

This is a useful step to explain the real processes of photon and actual processes in quantum chromodynamics QCD[17].

$$p\bar{p} \rightarrow \gamma + \gamma$$

The positron and proton have equal electric charge of $+1$, antiproton and electron have also equal charge of -1. Independent of their masses about color-charges symmetry we have:

$$k \triangleright = e^+, \qquad k \triangleleft = e^-$$

$$u = \frac{2}{3}k \triangleright, \qquad d = \frac{1}{3}k \triangleleft$$

$$\bar{u} = \frac{2}{3}k \triangleleft, \qquad \bar{d} = \frac{1}{3}k \triangleright$$

$$p\bar{p} = (uud)(\bar{u}\bar{u}\bar{d}) \rightarrow \gamma + \gamma$$

$$\left(\frac{2}{3}k \triangleright + \frac{2}{3}k \triangleright + \frac{1}{3}k \triangleleft\right) + \left(\frac{2}{3}k \triangleleft + \frac{2}{3}k \triangleleft + \frac{1}{3}k \triangleright\right) \Rightarrow \left(\frac{5}{3}k \triangleright + \frac{5}{3}k \triangleleft\right) = n(\triangleright + \triangleleft)$$

Therefore, there is color-charges symmetry in the output of process. By considering the structure of photon and by using the new definition of graviton, electric charge and charged particles, our point of view will change about the nature. In addition, this approach will provide us a new instrument to come over physical problems and conduct physics in a better path.

---

[17] - Freund A., "Exclusive annihilation p pbar to gamma gamma in a generalized parton picture" arXiv: 0208061v1, hep-ph, 2002

4

Space-time and Sub quantum energy

In this section with re-considering physical phenomena, according to a new definition of graviton that was presented in section one which by its using; the mechanism of graviton exchange between bodies/particle is described and surveyed.

Today the hottest problem in fundamental physics is quantum gravity. Quantum gravity was born in 1916, even before physicists had properly explained the other fundamental forces[1].

Gravitons are postulated because of the great success of quantum field theory (in particular, the Standard Model) at modeling the behavior of all other known forces of nature as being mediated by elementary particles: electromagnetism by the photon, the strong interaction by the gluons, and the weak interaction by the W and Z bosons. The hypothesis is that the gravitational interaction is likewise mediated by an elementary particle,

---

[1] - Gennady Gorelik, "Why Is Quantum Gravity So Hard?" SCIENTIFIC AMERICAN JOURNAl, 2011,

https://blogs.scientificamerican.com/guest-blog/why-is-quantum-gravity-so-hard-and-why-did-stalin-execute-the-man-who-pioneered-the-subject/

dubbed the graviton. In the classical limit, the theory would reduce to general relativity and conform to Newton's law of gravitation in the weak-field limit. However, attempts to extend the Standard Model with graviton has run into serious theoretical difficulties at high energies (see section 1, renormalization). Since classical general relativity and quantum mechanics are incompatible at such energies, from a theoretical point of view the present situation is not tenable. To find a solution to this problem, this section is based on the law of conservation of mass-energy, especial the conversion gravitational potential energy to electromagnetic energy and vice versa that was discussed in section one. Then according to this new approach, in this book, the mechanism graviton exchange between bodies/particles is described.

4.1 Energy curves space

General relativity is the geometric theory of gravitation and the current description of gravitation in modern physics. In general relativity, the universe has three dimensions of space and one of time and putting them together we get four dimensional spacetime, which gravity as an emergent effect from the spacetime curvature associated with distributions of energy. The central ideas of general relativity have been neatly summarized by John Archibald Wheeler[2]: "**matter tells space how to bend; space tells matter how to move**".

Energy cause space-time warp and curve and gravity is the exchange of gravitons. As the photon is packet of electromagnetic energy, a set of gravitons would be considered packets of the gravitational energy or space-time curvature. In spite of publishing many articles about graviton, but it has not been done any considerable work about mechanism of graviton exchange between bodies/particles. The reason is that the old graviton definition (in modern physics) is unable to describe this mechanism and also it is impossible to get the theory of the quantum gravity.

---

[2] - Relativity, space, time and gravity, The Physical World,
http://physicalworld.org/restless_universe/html/ru_4_24.html

Note that the aim of this section is to consider and describe the graviton exchange mechanism between the bodies. But the mechanism of graviton exchange between bodies must be compatible with the Newton's Second Law and the Universal Gravitational Law. In the other words, as graviton carrying gravitational force, it must be in accordance with the Newton's Second Law to change momentum, and consequently to change the energy of the bodies which causes the acceleration. On the other hand, as the exchange of graviton cause changing the momentum and energy, therefore there is a closely relation between gravitons (gravity) and photons (electromagnetic energy). That is why in this book, graviton is defined by using photon. On the other hand, virtual photons carry electromagnetic forces and the graviton carries the forces of gravity. It means there is a relations between gravity and electromagnetic forces. Therefore, in order to understand and describe the mechanism of graviton exchange between the bodies, it is necessary to describe the mechanism of virtual photons production based on the charged particles which was described in section two. So, we need review the general theory of relativity that is describes the gravity that usually is called space-time. Space-time is a mathematical model that combines space and time. There is two kind of space-times, Euclidean and non- Euclidean geometries. The geometry of curvature space-time is non-Euclidean geometry that has described in general relativity. To understand the space-time we have to look at how general relativity developed.

4.2 Problem of action at distance and speed of light

Aristotle believed that force could only be applied by contact; force at a distance being impossible[3]. Newton and other scientists had always been bothered that gravity appeared to act 'at a distance', a magical

---

[3] - O'Connor and Robertson, General relativity
http://www-groups.dcs.st-and.ac.uk/history/HistTopics/General_relativity.html

influence in empty space[4]. Newton's theory of gravitation was highly successful.

The early theories of electricity and magnetism were based on the Newtonian model of gravity, which is to say, they were based on the premise that isolated objects moving in the void of empty space exert forces on each other even when separated by some distance[5].

Maxwell's theory of electrodynamics, based on forces acting over insensible distances, proved to be tremendously successful[6]. Maxwell notes that there is a paradox caused by the attraction of like bodies. The energy of the medium must be decreased by the presence of the bodies and he said: "As I am unable to understand in what way a medium can possess such properties, I cannot go further in this direction in searching for the cause of gravitation"[7]. More precisely he required that the law governing propagation of gravitational action be Lorentz covariant and that the gravitational forces transform in the same way as electromagnetic forces[8]. Poincaré, in a paper in July 1905, suggested all forces should transform according the Lorentz transformations[9].

A long-standing problem in the study of the Solar System was that the orbit of Mercury did not behave as required by Newton's equations. Most of the effect is due to the pull from the other planets but there is a

---

[4] - Albert Einstein Cosmology, http://www.spaceandmotion.com/Physics-Albert-Einstein-Cosmology.htm

[5] - Why Maxwell Couldn't Explain Gravity, http://www.mathpages.com/home/kmath613/kmath613.htm

[6] - Last reference, why Maxwell couldn't explain gravity

[7] - O'Connor and Robertson, General relativity, http://www-groups.dcs.st-and.ac.uk/history/HistTopics/General_relativity.html

[8] - JOHN D. NORTON, EINSTEIN, NORDSTRÖM AND THE EARLY DEMISE OF SCALAR, LORENTZ COVARIANT THEORIES OF GRAVITATION, http://www.pitt.edu/~jdnorton/papers/Nordstroem.pdf

[9] - O'Connor and Robertson, General relativity, http://www-groups.dcs.st-and.ac.uk/history/HistTopics/General_relativity.html

measurable effect due to the corrections to Newton's theory predicted by the General Theory of Relativity[10].

4.3 Euclidean Relativity

In 1905–06 Henri Poincaré showed that by taking time to be an imaginary fourth space-time coordinate (ict), where c is the speed of light and i is the imaginary unit, a Lorentz transformation can formally be regarded as a rotation of coordinates in a four-dimensional space with three real coordinates representing space, and one imaginary coordinate representing time, as the fourth dimension[11].

Relativity theory traditionally uses the Minkowski hyperbolic framework. Euclidean relativity proposes a circular geometry as alternative that uses proper time (τ) as the fourth spatial dimension. The Euclidean metric is derived from the Minkowski metric by rewriting[12] as follow;

$$(cd\tau)^2 = (cdt)^2 - dx^2 - dy^2 - dz^2 \qquad (4\text{-}1)$$

Into the equivalent:

$$(cdt)^2 = dx^2 + dy^2 + dz^2 + (cd\tau)^2 \qquad (4\text{-}2)$$

The roles of time t and proper time τ have switched so that proper time τ becomes the coordinate for the 4th spatial dimension.

4.3.1 Manifold geometry and surface curvature

Manifold geometry: Manifold geometry is used in various fields of physics. In these geometries, spatial points and distances and the relation between them are shown algebraically. Thus, using the differential equation and vector analysis, geometric properties of the space can be studied.

[10] - Jose Wudka , Precession of the perihelion of Mercury, 1998, http://physics.ucr.edu/~wudka/Physics7/Notes_www/node98.html

[11] - Wikipedia, https://en.wikipedia.org/wiki/Minkowski_space

[12] - Euclidean Relativity , http://www.euclideanrelativity.com

Therefore, we deal with spatial coordinates in manifold geometries, the simplest of which is the Cartesian coordinate. The simplest form of manifold geometry is Euclidean manifold geometry (figure1).

Curvature of surface or Gaussian curvature: If we assume that the straight line is not curved, and we have to attribute a curvature number to it, this number should be $\kappa = 0$ and the curvature of a circle in radius r is $\kappa = \frac{1}{r}$.

Definition[13]: The osculating circle of a curve C at a given point P is the circle that has the same tangent as C at point P as well as the same curvature. Just as the tangent line is the line best approximating a curve at a point P, the osculating circle is the best circle that approximates the curve at P.

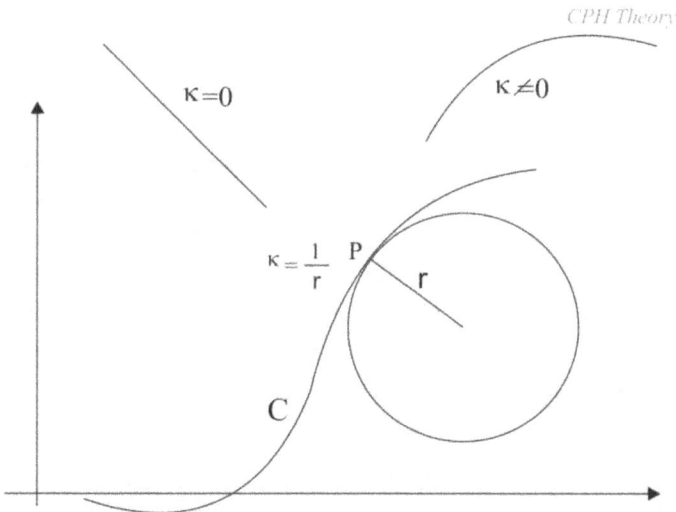

Fig4-1; The curvature of the curve and lines

To obtain the curvature of a curve at a point, the osculating circle is drawn at that point; the curvature of the curve at that point is equal to the

---

[13] - Wolfram MathWorld http://mathworld.wolfram.com/OsculatingCircle.htm

curvature of the osculating circle at that point. For the straight line the radius of the osculating circle is infinite at any point on it.

To determine the curvature of a surface at a point, we select two cross-sectional lines in the two main directions at that point and determine the curvature of these two lines in those points. Suppose the curvature of these two lines are κ_1, κ_2, then the curvature of the surface at that point is equal to the product of the multiplication of these two curves, namely:

$$\kappa = \kappa_1 \kappa_2 = \frac{1}{R_1 R_2} \qquad (4\text{-}3)$$

So, a sphere of radius r has Gaussian curvature $\frac{1}{r^2}$ everywhere, and a flat plane and a cylinder have Gaussian curvature 0 everywhere. The Gaussian curvature can also be negative, as in the case of a hyperboloid.

4.4 Non-Euclidean geometry

In 1854, Riemann delivered his monumental lecture "On the Hypotheses which lie at the Bases of Geometry"[14]. In this lecture, he described how to generalize Gauss' idea of surfaces and their curvatures to higher dimensions and thus singlehanded createdly the subject of Riemannian geometry[15]. Riemann discusses various possibilities by means of which an n-dimensional manifold may be endowed with a metric, and pays particular attention to a metric defined by the positive square root of a positive definite quadratic differential form.

In 1907, Einstein was preparing a review of special relativity when he suddenly wondered how Newtonian gravitation would have to be modified to fit in with special relativity[16]. Riemann's work was so ahead of his time

---

[14] - Bernhard Riemann, On the Hypotheses which lie at the Bases of Geometry, Translated by William Kingdon Clifford,
http://www.emis.de/classics/Riemann/WKCGeom.pdf

[15] - Peng Zhao, Riemannian Geometry and General Relativity, 2009,
http://www.damtp.cam.ac.uk/user/pz229/Teaching_files/GR.pdf

[16] - O'Connor and Robertson, General relativity,

that it did not receive its proper attention in the mathematical community until Einstein's formulation of general relativity in 1915[17]. Einstein imagined the spacetime as a geometric object whose curvature is determined by the distribution of energy and matter. Thus gravitational force is no longer a force in the Newtonian sense but a mere manifestation of the curvature of spacetime. The space-time is a manifold M and a space-time event is represented by a point $p \in M$. The worldline of the particle is traced by a curve in M. There are several basic tenets of general relativity.

• Space-time is a semi-Riemannian manifold.

• Free particles follow geodesics (figure 4- 2).

• Curvature tells matter how to move, and matter tells space-time how to curve[18].

Intrinsic properties - those that can be measured on the surface without regard to how it is embedded in an ambient space. In Euclidean geometry, the shortest distance between two points can be found using Pythagoras's theorem.

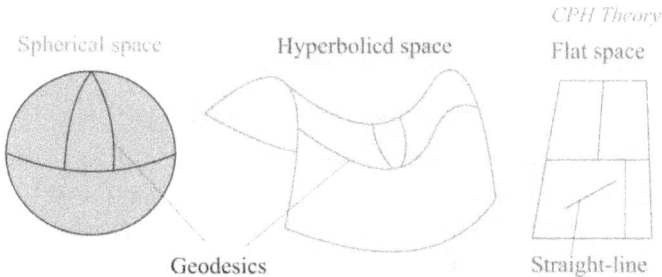

Fig;4- 2; Euclidean and non- Euclidean geometries

http://www-groups.dcs.st-and.ac.uk/history/HistTopics/General_relativity.html

[17] - Peng Zhao, last link

[18] - Peng Zhao, last link

What Riemann discovered was a more powerful, general form of Pythagoras's theorem that works on curved surfaces, even when the curvature is in more than two dimensions and varies from one place to another. In this looking-glass world of curved space, the familiar idea of distance is replaced by the broader concept of something called a metric, from the Greek for "measure," while curvature is similarly described by a more elaborate mathematical object. Gauss had found that the curvature in the neighborhood of a point of a specified two-dimensional geometry is given by a single number: the Gaussian curvature. Riemann showed that six numbers are needed to describe the curvature of a three-dimensional space at a given point, and that 20 numbers at each point are required for a four-dimensional geometry: the 20 independent components of the so-called Riemann curvature tensor[19].

4.5 Principle of Equivalence

Experiments performed in a uniformly accelerating reference frame with acceleration a are indistinguishable from the same experiments performed in a non-accelerating reference frame which is situated in a gravitational field where the acceleration of gravity = g = -a = intensity of gravity field"[20]. Before of Einstein, Hilbert had submitted a paper the foundations of physics which also contained the correct field equations for gravitation[21].

Principle of Equivalence has resulted:

1-   One way of stating this fundamental principle of general relativity is to say that gravitational mass is identical to inertial mass.

2-   One of the implications of the principle of equivalence is that since photons have momentum and therefore can be attributed an inertial mass, they must also have a gravitational mass. Thus photons should be deflected

---

[19] - General theory of relativity, Encyclopedia of Science, http://www.daviddarling.info/encyclopedia/G/genrel.html

[20] - Hyperphysics, Principle of Equivalence, http://hyperphysics.phy-astr.gsu.edu/hbase/Relativ/grel.html

[21] - Ivan T. Todorov, EINSTEIN AND HILBERT: THE CREATION OF GENERAL RELATIVITY, https://arxiv.org/pdf/physics/0504179.pdf

by gravity. They should also be impeded in their escape from a gravity field, leading to the gravitational red shift and the concept of a black hole. It also leads to gravitational lens effects.

3-   While attributing a kind of "effective mass" to the photon is one way to describe why the path of light is bent by a gravity field, Einstein's approach in general relativity is to associate a mass with a curvature of space-time, i.e. the existence of a mass will produce a curvature in space-time around it.

4-   From the point of view that light will follow the shortest path, or follows a geodesic of space-time, then if the Sun curves the space around it then light passing the Sun will follow that curvature[22].

How does the curvature effect matter? Matter always wants to travel in a straight line: it means the shortest line that joins two points. For a flat space that is straight line. But for a surface of a sphere that is curved, shortest line is geodesic[23].

4.5.1 Einstein field equation

The Einstein field equations is the set of 10 equations that describe the fundamental interaction of gravitation as a result of space-time being curved by mass and energy[24]. These equations are used to study phenomena such as gravitational waves[25]. The Einstein field equations (EFE) may be written in the form:

$$R_{\mu\nu} - \frac{1}{2}g_{\mu\nu}R = \frac{8\pi G}{c^4}T_{\mu\nu} \qquad (4\text{-}4)$$

Where $R_{\mu\nu}$ is the Ricci curvature tensor, R is the scalar curvature, $g_{\mu\nu}$ is the metric tensor, G is Newton's gravitational constant, c is the speed of light in vacuum, and $T_{\mu\nu}$ is the stress momentum-energy tensor.

---

[22] - Hyperphysics, last link

[23] - How can gravity act through empty space? Institute of Astronomy, http://www.ast.cam.ac.uk/public/ask/2580

[24] - Pinterest, https://www.pinterest.com/pin/178032991497358424

[25] - LIGO Caltech, https://www.ligo.caltech.edu/page/what-are-gw

Ricci curvature is the mathematical object[26] that controls the growth rate of the volume of metric balls in a manifold.

Scalar curvature [27]of a Riemannian manifold is given by the trace of the Ricci curvature tensor.

Metric tensor[28], g_{ij} is a function which tells how to compute the distance between any two points in a given space. Its components can be viewed as multiplication factors which must be placed in front of the differential displacements dxi in a generalized Pythagorean Theorem:

$$ds^2 = g_{11}dx_1^2 + g_{12}dx_1dx_2 + g_{22}dx_2^2 + \cdots \quad (4\text{-}5)$$

Stress–energy tensor in local coordinates, the stress-energy tensor may be regarded as a 4x4 matrix T_{ab} at each point of space-time.

The Einstein field equations are not a dynamical equations that describe how matter and energy change the geometry of space-time, this curved geometry being interpreted as the gravitational field of the matter source. Einstein tried to propound geometrical structures of space by mathematical equations. So, he used non-Euclidian geometry. There are three considerable notes on Einstein's equations;

1- Einstein Field Equations do not come from the equivalence principle directly. These equations are simply equations that are suitable for general relativity.

2- There is a physical explanation for the path of light in a gravitational field. Although explaining the frames of reference is a physical concept, there is not any explanation of how gravitational field affects photons in general relativity. Then how we can explain this phenomenon by quantum mechanics?

---

[26] - Ricci Curvature Tensor, Wolfram MathWorld,
http://mathworld.wolfram.com/RicciCurvatureTensor.html

[27] - MathOverflow, https://mathoverflow.net/questions/30035/some-questions-about-scalar-curvature

[28] - Metric Tensor, Wolfram MathWorld, http://mathworld.wolfram.com/MetricTensor.html

3-   Space-time is a continuous quantity in general relativity. But the changing of photon frequency and production of energy are quantized. That gravitational blueshift (or redshift) is a special case of gravitational field that affects the photon.

General relativity spacetime has no intrinsic properties other than its curved geometry[29]. At large scales, gravity just doesn't behave the way Einstein's theory predicts[30]. We need solve quantum gravity problem, it shows how to modify Einstein Field equation. Problem is not of Equivalence', in fact problem is that Einstein Field Equations do not come from the equivalence principle directly. To formulate a theory of quantum gravity, the first step is to describe the graviton exchange mechanism between particles.

4.6 Graviton Exchange Mechanism

According to the previous section, the charged particles are producing positive and negative color-charge. Negative charged particles absorb positive color-charges G^+ and produces and emits positive virtual photons γ^+, and repels negative color charges G^-. Also positive charged particle absorbs negative color-charges G^- and produces and emits negative virtual photons γ^-, and repels positive color charges G^+. In addition, according to the law of conservation of electric charge, amount and number of positive and negative electric charges are equal, therefore, the amount and number of positive and negative color-charges is released in space are equal. So, space is full of positive and negative color-charges and normal graviton G, G^+, G^-, they are released faster than the speed of light in

[29] - Erik P. Verlinde, Emergent Gravity and the Dark Universe, Arxiv, 2016, https://arxiv.org/pdf/1611.02269.pdf

[30] - New theory of gravity might explain dark matter, Phys.org, 2016, https://phys.org/news/2016-11-theory-gravity-dark.html

space. Density of G , G^+ , G^- around objects is proportional to $1/_{r^2}$, where r is the distance of the body.

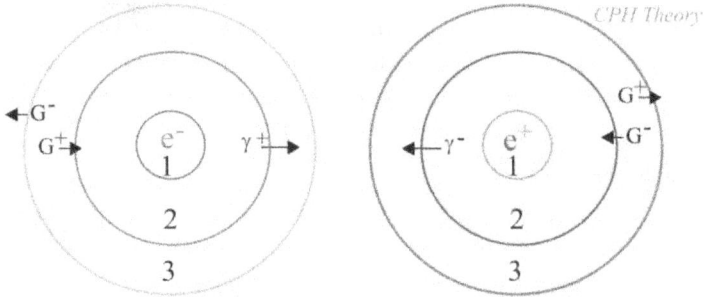

A.   Electron; Area3, G convert to G^- , G^+, then G^- moves to far and G^+ moves to area2

Area2: Spinning electron, magnetic field compacts G^+s and repels virtual positive photon γ^+ .

B.   Positron; Area3, G convert to G^- , G^+, then G^+ moves to far and G^- moves to area2

Area2: Spinning positron, magnetic field compacts G^-s and repels virtual negative photon γ^- .

Fig4- 3; area 2 around charged particles

According to the concepts of color-charges and magnetic-colors G^- , G^+ , G_m^+ , G_m^- and negative and positive virtual photons γ^+ , γ^- , it can be concluded that the Dirac equation is not only for the high-energy photon and pair production electron-positron, in fact Dirac Sea can be generalized to all of space (section 2). This means all objects / particles in the universe, are immersed in the Dirac Sea. The amount of production of γ^+ , γ^-, and their energies depend to density of graviton (in fact, density of G^- , G^+ , G_m^+ , G_m^- , G) in space and around the particles. As seen in figure (4- 3), area 2 around any charged particle is the birthplace of virtual photons that carry the electromagnetic force.

But area 3 around any charged particle is the birthplace of sub quantum energies that carry the gravity force. Positive and negative sub quantum energies $\triangleright$ and $\triangleleft$ are created that make the gravitational potential energy (figure 4- 4). By comparing the ratio of the electromagnetic force and gravitational force between two electrons[31], as well as the energy of sub quantum energy $\triangleright$, $\triangleleft$ and energy of virtual photon γ^+, γ^- can be compared with each other.

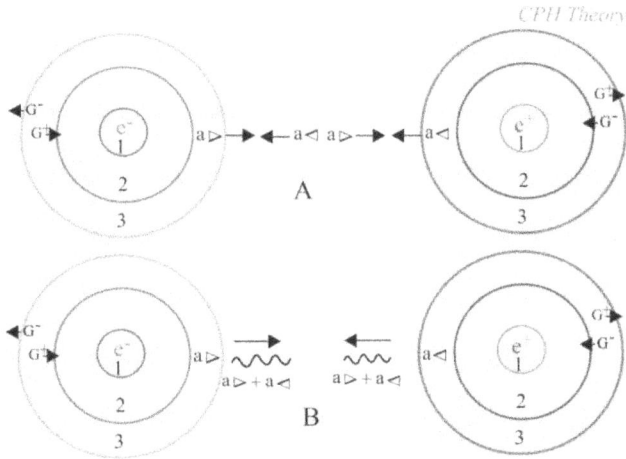

A.   Area 3; charged particles produce sub quantum energies $\triangleright$ and $\triangleleft$
B.   Area 3; sub quantum energies combine, gravitation energies are produced and particles attract each other

Fig4- 4; area 3 around charged particles

Example 1: The stone starts out with a certain amount of kinetic energy, but as it climbs it slows down and its kinetic energy decreases. Its kinetic energy convert to sub quantum energies $\triangleright$, $\triangleleft$, and sub quantum energies convert to color-charges and magnetic colors and normal graviton $G^-, G^+, G_m^+, G_m^-, G$ (similar to gravitational red-shift). Until, at the top of its flight, the stone is momentarily at rest. On the way down, first gravitons convert to color-charges and magnetic-colors G^-, G^+, G_m^+, G_m^- , and they

---

[31] - The electromagnetic repulsion between the electrons is stronger than their gravitational attraction by a factor of over 10^{42}!

convert to sub quantum energies ▷ , ◁ , then ▷ , ◁ combine and do form the kinetic energy. And in the end, in the collision stone with Earth, sparks and heat are generated.

Example 2: According to CPH Theory, gravity is a currency among between the objects. Consider the interaction between the Earth and the Moon: when a set of gravitons reach the Earth, convert to G^- , G^+, G_m^+ , G_m^- and combine with a set of G^- , G^+, G_m^+ , G_m^- from Earth, they do form binding energy between Moon and Earth. Same process happens around the Moon. These binding energies push the Earth and Moon toward each other. Because as to maintain equality times - positive and negative color-charges, there is a fixed ratio between the mass and the number of gravitons surrounding. When a set of gravitons reach the Moon, the other set move toward the Earth and push the Moon toward the Earth. So Earth (in fact everything) is bombarded by gravitons continuously. Due to the fact that everything is made up of sub quantum energy, the classical concept of acceleration and relativistic Newton's second law needs to be reviewed (see section 5).

Generation and absorption of the gravitational energy by two bodies which are located in their gravitational fields of each other are done by interaction between their gravitational fields, too. According to Newton's universal gravitation law, the gravitational force between two objects is directly proportional to the product of their masses. Then, if we describe any gravitational interaction between two or more atoms/subatomic particles, we can generalize and extend this description to large bodies such as stars and even the galaxies.

According to the above description, both charged particles and neutral particles are producing and exchanging gravitational energy by exchanging gravitons. This can be called gravitational binding energy. Gravitational binding energy causes atoms and molecules attract each other and it makes large bodies such as planets, stars and galaxies. Neutral particles such as photons are attracted by exchanging gravitons with other particles, too, because neutral particles are made up of positive and negative sub quantum energies. In structure of atom, each particle is linked with its adjacent particle by gravitational binding energy. Moreover, it is related by releasing the color-charges in the space. Each atom is linked with neighboring atoms by gravitational binding energy and by releasing color-charges; it shows its presence (figure 4- 5). Thus, large bodies such as stars and galaxies are created.

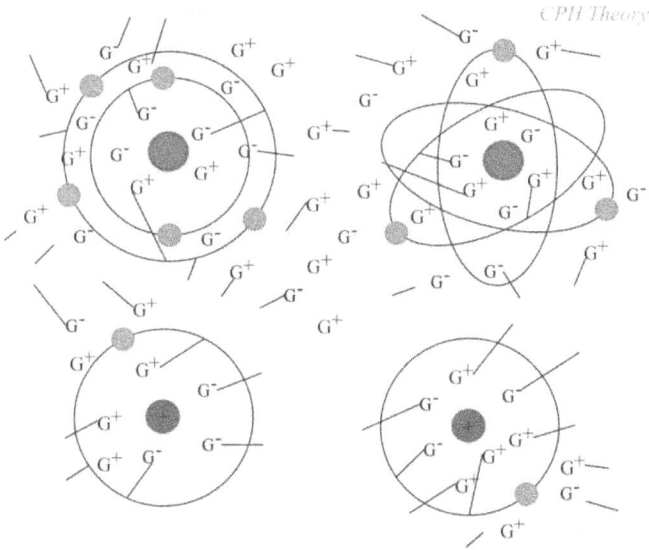

Fig4- 5: Each atom has a gravitational binding energy with the surrounding atoms and it is related with the space by releasing color-charges.

4.7 What is space-time, really?

In general relativity, space-time is a mathematical model that combines space and time into a single interwoven continuum. Minkowski said[32]: "Henceforth, space by itself, and time by itself, are doomed to fade away into mere shadows, and only a kind of union of the two will preserve an independent reality." According to the general relativity, matter curves the space-time around itself. So, path of light is bent in gravitational field.

---

[32] - Peter Russell, http://www.peterrussell.com/SG/Ch5.php

All our theories today seem to imply that the universe should contain a tremendous concentration of energy, even in the emptiest regions of space[33]. A ball spinning in a vacuum should never slow down, since no outside forces are acting on it. But over time, a spinning object will gradually slow down, even if equal numbers of virtual photons bombard it from all sides[34]. It shows vacuum has friction. So we have to change our point of view about space, time and energy.

Select an object. It is not important that what we have chosen, how we look at this physical existence is important. I choose a ball that is on the surface of room. I shoot the ball outside of room. What happens for ball's space? Nothing, ball and its space moves from room's space to salon. In fact space of ball moved with it. We cannot take space from the body or body from the space, object and space are intertwined with each other. Fill a glass with water, empty it and refill it with syrup. In this process; water with its space leaves the glass and syrup with its space enters into glass. We can never separate the physical existence from its space. In fact space is a part of physical existence. Also, time and physical existence are not independent. Ball is a physical existence; it is made up of some molecules with their spaces at t (time relative to other clock). Ball is a clock as the same as other physical existences. There are many atoms in the structure of ball, how can we prevent movement of subatomic particle in the ball? Never. Generally, mass (or energy), space, motion and time are intertwined and inseparable from each other.

Space is full of gravitons, which we know that the gravitational potential energy. Although the origin of mass is gravitational energy, but gravitational energy is a physical entity. However, we cannot review gravitational energy independent of own space, time and motion. That is why the light path is curved in space-time.

Such as a lens consists of a mass M which deflects light from a distant light source S in such a way that the light reaches an observer O along two

---

[33] - Gordon Kane, The dawn of physics beyond the standard model,
https://experts.umich.edu/en/publications/the-dawn-of-physics-beyond-the-standard-model
[34] - David Harris, Vacuum has friction after all, 2011,
https://www.newscientist.com/article/mg20927994-100-vacuum-has-friction-after-all/

different paths[35]. As a consequence, O will see two distinct images of S (figure 4- 6). The bending of space-time, must be explained directly by the gravitational potential energy. Describing curved space-time is as a result of interaction between graviton and photon. It is described here by binding energy between a massive body and two same photons. There is two beams of light in figure (4- 6). There is two beams of light in figure (4- 6). Suppose there are two same photons (1 and 2) in these beams. Simultaneous, they reach at line 1. Consider to photon 2 (right side of figure). It has energy $E_1 = a_1(\triangleright + \triangleleft)$ at line 1, by gravitational blueshift, a lot color – charges and magnetic colors enter to structure of photon one, at line 2, its energy reaches to $E_2 = a_2(\triangleright + \triangleleft)$ so that $E_2 > E_1$ *because* $a_2 > a_1$.

Fig4- 6; paths of two same photons around a massive body

---

[35] - Spotlights on relativity, http://www.einstein-online.info/spotlights/grav_lensing_history

As we know intensity of gravitation potential depends to distance of body, and gravitational potential energy (in fact, gravitons) pushes photon toward the massive body and bends the path of photon. Photon shifts to red between line 2 and line 3, and at line 3 it has energy equal $E_1 = a_1 (\triangleright + \triangleleft)$. These blueshift and redshift cause the observer does observe S_1 and S_2 instead of one source S.

In the recent decades, quantum gravity has been the biggest problem in physics. This problem caused physicists to work on two theories in quantum mechanics and general relativity to be compatible by presenting mathematical equations. But in CPH Theory it has been working directly with reviewing and analysis of the physical phenomena such as structure of elementary particles and their interaction to be described. That is why the graviton exchange mechanism between bodies according to structure of particles have been reviewed and analyzed.

We can use this approach for unifying quantum fields. By summarizing main idea of CPH Theory about birthplace of fundamental bosons, integration will be very easy than that someone can thought (figure 4- 7). Take a look at figures (2- 3, 3- 2 and 4- 3). The important parts of them can be seen in figure (4- 7).

Today, there is no way to explain the process that describes how particles produce exchange particles in modern physics. We can definitely say that the best way for unifying the interactions is generalizing interaction between charged particles to photon structure and vice versa. This new view on photon means that we can redefine the graviton and electromagnetic energy. Electromagnetic energy converts to matter and anti-matter such as charged particles. Charged particles use gravitons and generate electromagnetic field. This way of looking at the problem shows how two same charged particles repel each other in far distance and absorb each other at a very small distance .

With a detailed look at the sub-quantum space, we can investigate better the interaction between quarks in a very small space of proton. Using such approach to generate matter-anti matter, we can explain that how bosons are generated from fermions and then we can provide a mechanism for the unification of forces.

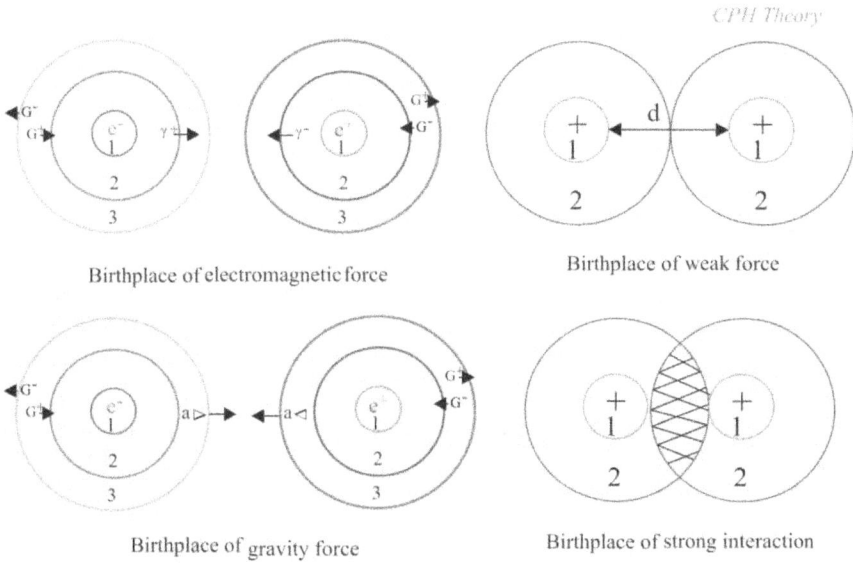

Birthplace of electromagnetic force

Birthplace of weak force

Birthplace of gravity force

Birthplace of strong interaction

Fig4- 7: the birthplaces of fundamental interactions

We can use the sub-quantum space to describe the nature of time in order to understand better the nature of space-time, and review of thermodynamics laws and entropy (see section 6). As long as we do not review relativistic Newton's second law, physics does not stand on its actual position. Moreover, by reconsidering the relativistic Newton's second law, it can be explained the expansion of the universe better and more real than the past (see next section).

5

Cosmological equations and Sub quantum energy

Motion is an intrinsic property of physical existence. But there is a problem about concept of acceleration in theoretical physics.

First of all, what we know about acceleration? And what is the definition of acceleration? In physics, acceleration is the rate at which the velocity of an object changes over time that in classical mechanics is given by Newton's second law with $a = F/m$. In special relativity an accelerating particle has a world-line which is not straight. This is not difficult to handle. It's a common misconception that special relativity cannot handle accelerating objects or accelerating reference frames. Sometimes it's claimed that general relativity is required for these situations, the reason being given that special relativity only applies to inertial frames. This is not true. Special relativity treats accelerating frames differently from inertial frames, but can still deal with accelerating frames. And accelerating objects can be dealt with without even calling upon accelerating frames[1].

At the beginning of the 20th century, Newton's second law was corrected considering the limit of speed c and the relativistic mass. At that

---

[1] - Philip Gibbs, "Can Special Relativity Handle Acceleration?" 1996, [Online] available; http://math.ucr.edu/home/baez/physics/Relativity/SR/acceleration.html

time there has not been a clear understanding of the subatomic particles and basically there was little research in high energy physics .

However, Newton and Einstein define the acceleration regardless of the structure of particles (in classical mechanics and relativity). This definition belongs to Newton era or macroscopic level. It should be noted that the interaction between large objects (e.g. collision of two bodies) under the action of the quantum layer (in fact sub quantum level) has been done. Thus, according to quantum mechanics and mass–energy equivalence $E = mc^2$, we must redefine acceleration.

5.1 Reconsidering the relativistic Newton's second law

In Newtonian mechanics, the time is absolute, and infinite speed was accepted and Newton's second law with constant mass that was presented as follows:

$$F = \frac{dp}{dt} = m\frac{dv}{dt} \qquad (5\text{-}1)$$

By proposing of the relativity and limit speed of light c, the equation (5-1) was corrected so that the limitation of speed must had applied. So the Newton's second law was as follows:

$$F = \frac{dp}{dt} = \frac{d(mv)}{dt} = v\frac{dm}{dt} + m\frac{dv}{dt} \qquad (5\text{-}2)$$

Bucherer by measuring the ration of charge with respect to mass of electron e/m in different speed, showed that mass increases along with increasing the speed[2]. Bucherer experiment was an experimental verification of relativistic mass and due to accuracy of relativistic Newton's second law (equation 5- 2). Increasing the mass of electron while passing from accelerator tunnel (imposing external force) is due to obtaining energy and energy has mass. The subject that an object (or a particle) cannot move

---

[2] - Thos. Lewis, "The Interpretation of the Results of Bucherer's Experiments on e/m" Proceedings of the Royal Society of London. Series A, Containing Papers of a Mathematical and Physical Character Vol. 107, No. 743 (Mar. 2, 1925), pp. 544-560

with speed of light, is due to structure of matter and mechanism of interaction of field with matter that by principle of graviton and sub quantum energy, being constant of the value of speed can be generalized from energy to mass. Therefore, it is worthy to reconsider Bucherer experiment. In Bucherer experiment, consider an electron with mass m_0, speed v_1 and at moment t_1 is moving on direction of an axis, accelerates under effect of force F and at the moment t_2, its speed is v. In time interval $\Delta t = t_2 - t_1$, electron gains energy equal to E , and its mass increases as $m_E = nm_{SQE}$ (equation 1- 34). So, we can write:

At moment t_1, speed and momentum of electron is v_1 and m_0v_1 respectively.

At moment t_2: speed and momentum of electron is v and mv respectively. So that:

$$m = m_0 + m_E , m_E = nm_{SQE} \Rightarrow m = m_0 + nm_{SQE}$$

According to the conservation law of linear momentum, the momentum of input electron m_0v_1 plus momentum of energy gained in interval time $\Delta t = t_2 - t_1$, must be equal to output momentum. Therefore, we have:

$$m_0v_1 + nm_{SQE}c = mv \qquad (5\text{-}3)$$

But according to equation (5- 3), $m = m_0 + nm_{SQE}$ and by replacing, we have:

$$m_0v_1 + nm_{SQE}c = (m_0 + nm_{SQE})v$$

$$v = \frac{m_0v_1 + nm_{SQE}c}{m_0 + nm_{SQE}} < c \qquad (5\text{-}4)$$

Because v_1 is the speed of input electron at t_1 and all laboratory experiments show that the speed of the electron is less than the speed of light c, also, if the electron speed was greater than the speed of light, there was no need to relativity. So, according to $v_1 < c$, we have:

$$v_1 < c \therefore v = \frac{m_0 v_1 + n m_{SQE} c}{m_0 + n m_{SQE}} < \frac{m_0 c + n m_{SQE} c}{m_0 + n m_{SQE}} = c \Rightarrow v < c$$

In Newton's second law, the extra mass can be related to gained energy. So, we have:

$$\frac{dm}{dt} = \frac{n m_{SQE}}{dt} = \frac{1}{c^2} \frac{dE}{dt}$$

$$F = \pm \frac{v}{c^2} \frac{dE}{dt} + m \frac{dv}{dt} \qquad (5\text{-}5)$$

The $\pm$ sign in equation (5- 5) has been marked for two states of the increasing and decreasing energy (collinear or non-collinear directional variations in force and speed). The Newton's second law in equation (5- 5) increases our ability to have better cognition and explanation of physical phenomena. With such an approach to physical and astrophysical events, the explanation of universe will be more real. According to the principle of sub quantum energy SQE, the value of speed of formed particles is continuously constant in subatomic particles and external force just can convert the motion of SQEs from linear motion to nonlinear motion and vice versa.

Thus, the speed of light is constant in vacuum, but it changes in other mediums such as air or water and as soon as entered into vacuum, it moves with the same constant speed. Speed of other subatomic particles is a function of internal reactions between particles (interaction between SQEs in the structure of particles). In fact, in Boucherer experiment, electron gains energy while accelerating and after exiting from accelerator tunnel loses its energy due to striking with other particle or passing through a field that gives it a negative acceleration and comebacks to the earlier state regarding the mass.

This experiment can be used continuously to verify relativity, but it cannot be explained real interaction between force and mass with relativistic mass. The reason for limitation of speed must be investigated in the structure of matter. Moreover, the relativistic attitude is a post structural view and explains observer's point of view and it has not paid attention to the inherent nature of phenomena. While CPH Theory tried to investigate about physical phenomena regarding the internal structure of particles and explain the necessity of reconsidering in relativistic Newton's second law.

In fact, the external force cannot change the value of speed under any physical conditions and it just can convert linear motion of formed particles of matter and energy to nonlinear motion and vice versa. With such insight, we can have better cognition from singularity by definition of absolute black hole, explain before big bang and express the reason of inflation theory.

5.2 Do physical laws break down?

A gravitational singularity or space-time singularity is a location where the quantities that are used to measure the gravitational field become infinite in a way that does not depend on the coordinate system[3]. Hawking said: "the Big Bang, all the matter in the universe, would have been on top of itself. The density would have been infinite. It would have been what is called, a singularity. At a singularity, all the laws of physics would have broken down. This means that the state of the universe, after the Big Bang, will not depend on anything that may have happened before, because the deterministic laws that govern the universe will break down in the Big Bang. The universe will evolve from the Big Bang, completely independently of what it was like before. Even the amount of matter in the universe, can be different to what it was before the Big Bang, as the Law of Conservation of Matter, will break down at the Big Bang"[4].

It means there were two physical laws are ruled the universe, before and after the Big Bang. I disagree with it. There are various theories in physics, but nature is unique. This is not nature's problem that we have various theories; nature obeys simple and unique law. So, we should improve our theories. It should be noted that the laws of physics are not independent of the physical nature of time. Understanding the physical nature of time, without understanding the structure of elementary particles is impossible. So, the next section is dedicated to the physical nature of time.

However, Hawking later revised this to claim that general relativity breaks down at times prior to the Big Bang, and hence no singularity could

---

[3] - Matt Williams, Universe Today, 2017, https://www.universetoday.com/84147/singularity/

[4] - S. Hawking, The Beginning of Time, http://www.hawking.org.uk/the-beginning-of-time.html

be predicted by it[5]. Not only general relativity is facing a problem, all attempts to treat gravity simply like one more quantum force have failed. The problem of gravity, is a historical problem.

Since Newton introduced his universal gravitational law, although this law is described in celestial mechanics was a dazzling success, but was associated with some questions and ambiguities. Newton found by the of gravity, stars shall attract each other and thus do not seem to be at rest. In 1692, Newton wrote in a letter to Richard Bentley: "As to your first query, it seems to me that if the matter of our sun and planets and all the matter in the universe were evenly scattered throughout all the heavens, and every particle had an innate gravity toward all the rest, and the whole space throughout which this matter was scattered was but finite, the matter on the outside of the space would, by its gravity, tend toward all the matter on the inside, and consequently, it falls down into the middle of the whole space and there compose one great spherical mass. But if the matter was evenly disposed throughout an infinite space, it could never convene into one mass; but some of it would convene into one mass and some into another, so as to make an infinite number of great masses, scattered at great distances from one to another throughout all that infinite space.[6] "

The next problem is that according to Newton's gravity law a body could attracts unlimited other objects and it grows to infinity, this is also unrealistic, because it is not compatible with experience. In generally, how we can resolve infinity problem of physics? My question is, if the universe collapses, will it reach to infinite density and an infinitely small space? Or is there a force that will counteract it? In this section, the answer to this question has been given by a new definition of singularity. In fact, the theme of this section is removing infinities of physics.

---

[5] - Last link of Matt Williams, Universe Today

[6] - Joseph Joubert, Pensdes, "The Newton-Bentley Exchange"
https://ned.ipac.caltech.edu/level5/Sept02/Saslaw/Saslaw1_2.htm

5.3 Cosmological models of the universe

A static universe, is a cosmological model in which the universe is both spatially infinite and temporally infinite, and space is neither expanding nor contracting. Such a universe does not have spatial curvature; that is to say that it is 'flat'. A static infinite universe was first proposed by Giordano Bruno: "The universe is then one, infinite, immobile.... It is not capable of comprehension and therefore is endless and limitless, and to that extent infinite and indeterminable, and consequently immobile"[7]. In contrast to this model, Albert Einstein proposed a temporally infinite but spatially finite model as his preferred cosmology in 1917, in his paper Cosmological Considerations in the General Theory of Relativity[8].

To derive his 1917 cosmological model, Einstein made three assumptions that lay outside the scope of his equations[9].

1- The universe is homogeneous and isotropic in the large (i.e., the same everywhere on average at any instant in time).

2- The total volume of a three-dimensional space with uniform positive curvature would be finite but possess no edges or boundaries (to be consistent with the first assumption).

3- The universe as a whole is static - i.e., its large-scale properties do not vary with time.

When Einstein first studied the universe at large using the General Theory of Relativity he discovered that his equations predicted a universe which was either expanding or contracting and this was contradicted with the best astronomical observations at the time. He then modified his equations to satisfy the observations. This modification corresponds to the assumption that the whole universe is permeated with a constant pressure

---

[7] - Giordano Bruno, Teofilo, in Cause, Principle, and Unity, "Fifth Dialogue," (1588), ed. and trans. by Jack Lindsay (1962)

[8] - O'Raifeartaigh C. and McCann B." Einstein's cosmic model of 1931 revisited: an analysis and translation of a forgotten model of the universe",
arXiv:1312.2192v2 [physics.hist-ph], 2014

[9] - The Cosmological Constant,
http://abyss.uoregon.edu/~js/glossary/cosmological_constant.html

(which in his case balanced the expansion yielding a steady universe). This universal pressure is called the cosmological constant Λ (lambda).

Einstein's static universe is closed and contains uniform dust and a positive cosmological constant with value $\Lambda_E = 4\pi G\rho/c^2$, where G is gravitational constant, ρ is the energy density of the matter in the universe and c is the speed of light[10]. The **radius of curvature** of space of the Einstein universe is equal to:

$$R_E = \frac{c}{\sqrt{4\pi G\rho}} \qquad (5\text{-}6)$$

The cosmological constant appears in Einstein's field equation in the form of:

$$R_{\mu v} - \frac{1}{2} g_{\mu v} R + \Lambda g_{\mu v} = \frac{8\pi G}{c^4} T_{\mu v} \qquad (5\text{-}7)$$

Greek capital letter lambda Λ is the cosmological constant. Cosmological constant is the value of the energy density of the vacuum of space.

Where $R_{\mu v}$ is the Ricci curvature tensor, R is the scalar curvature, $g_{\mu v}$ is the metric tensor, Λ is the cosmological constant, G is Newton's gravitational constant, c is the speed of light in vacuum, and $T_{\mu v}$ is the stress–energy tensor.

5.4 Friedmann Equation

At almost exactly the same time, Friedmann carefully revised the Einstein's cosmological equations and he published his classic relativistic cosmology. His key realization was that isotropic world models had to have isotropic curvature everywhere. In the paper of 1922, Friedmann found the solutions for expanding the universe models with closed spatial geometries, including those that expand to a maximum radius and then collapse to a

---

[10] - Domingos Soares, Einstein's static universe, 2012, http://arxiv.org/pdf/1203.4513.pdf
Cosmology , http://www.astronomy.ohio-state.edu/~dhw/A873/notes3.pdf

singularity. Friedmann showed that there exist expanding solutions that are unbounded with hyperbolic geometry[11]. The differential equations[12] that he derived were:

$$\left[\left(\frac{1}{R}\frac{dR}{dt}\right)^2 - \frac{8}{3}\pi G\rho\right] R^2 = -kc^2 \qquad (5\text{-}8)$$

After Hubble discoveries on the universe's expansion, Friedmann's equation was as follows:

$$\left(H^2 - \frac{8}{3}\pi G\rho\right) R^2 = -kc^2 \qquad (5\text{-}9)$$

Where $H = \left(\frac{1}{R}\right)\frac{dR}{dt}$ is Hubble "constant", G is the gravitational constant, ρ is the universe mass density, c the speed of light and the parameter k is 0 for Euclidean Geometry or flat space, +1 , elliptic space and -1, hyperbolic space. One can write $\rho = \rho_0(R_0/R)^3$, where ρ_0 and R_0 are the present day values of the density and radius of the universe.

In other words, Friedmann raised the possibility of a dynamic universe, which changes in size over time. In fact, Friedmann introduced the expression "expanding universe." Moreover, one of his solutions modeled a cosmos which began in a singularity – an infinitesimally small point. It even had an expansion rate which increased over time, just as modern observations indicate. Einstein wrote a short note in the German Physics Journal, Zeitschrift fur Physik, calling Friedmann's non-stationary world "suspicious." Friedmann immediately sent the great physicist an extended letter detailing his work. Six months later, Einstein wrote in the journal:". .

---

[11] - MALCOLM S. LONGAIR, "A Brief History of Cosmology", Carnegie Observatories Astrophysics Series, Vol.2 Measuring and Modeling the Universe, 2004, ed. W. L. Freedman (Cambridge: Cambridge Univ. Press)

http://www.astro.caltech.edu/~george/ay21/readings/longair.pdf

[12] - The Friedmann Equation,

https://ned.ipac.caltech.edu/level5/Peacock/Peacock3_2.html

http://hyperphysics.phy-astr.gsu.edu/hbase/astro/fried.html

my criticism . . . was based on an error in my calculations. I consider that Mr. Friedmann's results are correct and shed new light."[13]

Einstein was arguing against the solutions describing expanding universe. It appears that the community accepted the concept of an expanding universe largely due to the work of Lemaitre. By 1931, Einstein himself had rejected the cosmological term as superfluous and unjustified. There is no direct record that Einstein ever called cosmological constant his biggest blunder. It is possible that this often repeated "quote" arises from Gamow's recollection: "When I was discussing cosmological problems with Einstein, he remarked that the introduction of the cosmological term was the biggest blunder he ever made in his life"[14].

5.5 Big Bang Theory

In 1927, Lemaître proposed an expanding model for the Universe to explain the observed redshifts of spiral nebulae, and forecast the Hubble law. He based his theory on the work of Einstein and De Sitter, and independently derived Friedmann's equations for an expanding universe. In 1929, Edwin Hubble provided a comprehensive observational foundation for Lemaître's theory. Hubble's experimental observations discovered that, relative to the Earth and all other observed bodies, galaxies are receding in every direction at velocities (calculated from their observed red-shifts) directly proportional to their distance from the Earth and each other. In 1929, Hubble and Milton Humason formulated the empirical redshift distance law of galaxies, nowadays known as Hubble's law, which, once the redshift is interpreted as a measure of recession speed, is consistent with the solutions of Einstein's general relativity equations for a homogeneous, isotropic expanding space. The isotropic nature of the expansion was direct proof that it was the space (the fabric of existence) itself that was expanding, not the bodies in space that were simply moving further

---

[13] - Mark Egdall, "Alexander Friedmann: Unsung Hero of Modern Cosmology", 2012, http://www.decodedscience.org/alexander-friedmann-unsung-hero-of-modern-cosmology/19423

[14] - A brief history of cosmological constant, [Online] available; http://ned.ipac.caltech.edu/level5/Sept02/Padmanabhan/Pad1_2.html

outward and apart into an infinitely larger preexisting empty void. It was this interpretation that led to the concept of the expanding universe. The law states that the greater the distance between any two galaxies, the greater their relative speeds of separation. This discovery later resulted in the formulation of the Big Bang model.

The Big Bang theory is an effort to explain what happened at the very beginning of our universe. Discoveries in astronomy and physics have shown beyond a reasonable doubt that our universe did in fact have a beginning. Prior to that moment there was nothing; during and after that moment there was something: our universe. The big bang theory is an effort to explain what happened during and after that moment. According to the standard theory of Big Bang, our universe sprang into existence as "singularity" around 13.7 billion years ago. What is a "singularity" and where does it come from? Well, to be honest, we don't know for sure. Singularities are zones which defy our current understanding of physics. They are thought to exist at the core of "black holes." Black holes are areas of intense gravitational pressure. The pressure is thought to be so intense that finite matter is actually squished into infinite density (a mathematical concept which truly boggles the mind). These zones of infinite density are called "singularities." Our universe is thought to have begun as an infinite small, infinitely hot, infinitely dense, something - a singularity. Where did it come from? We don't know. Why did it appear? We don't know[15].

After its initial appearance, it apparently inflated (the "Big Bang"), expanded and cooled, going from very, very small and very, very hot, to the size and temperature of our current universe. It continues to expand and cool to this day and we are inside of it: incredible creatures living on a unique planet, circling a beautiful star clustered together with several hundred billion other stars in a galaxy soaring through the cosmos, all of which is inside of an expanding universe that began as an infinitesimal singularity which appeared out of nowhere for reasons unknown. This is the Big Bang theory[16]. Despite its successes, the standard big bang theory was too simple to be complete. The Inflation Theory proposes a period of extremely rapid (exponential) expansion of the universe during its first few

---

[15] - All about science, Big Bang Theory - An Overview, http://www.big-bang-theory.com/

[16] - Big Bang Theory - An Overview, AllAboutScience.org, http://www.big-bang-theory.com/default.htm

moments. Cosmologists introduced this idea in 1981 to solve several important problems in cosmology[17]. It was developed to explain several puzzles with the standard Big Bang theory, in which the universe expands relatively gradually throughout its history[18].

5.5.1 Problems with the Big Bang Theory

There is many problems about big bang theory such as[19]:

1- It violates the first law of thermodynamics, which says you can't create or destroy matter or energy.

2- Some critics say that the formation of stars and galaxies violates the law of entropy, which suggests systems of change become less organized over time.

3- Some astrophysicists and cosmologists argue that scientists have misinterpreted evidence like the redshift of celestial bodies and the cosmic microwave background radiation.

4- The early inflationary period of the big bang appears to violate the rule that nothing can travel faster than the speed of light.

5.6 The Theory of Inflation

In the corresponding theory of inflation, the Universe, because of properties of elementary particles not accounted for in the standard big bang models, expands for a fleeting instant at its beginning at a much higher rate than that expected for the big bang. This period, which is called the inflationary epoch, is a consequence of the nuclear force breaking away from the weak and electromagnetic forces that it was unified with at higher temperatures in what is called a phase transition.

---

[17] - The inflationary Universe, http://www.ctc.cam.ac.uk/outreach/origins/inflation_zero.php

[18] - Kane, G, "The dawn of physics beyond the standard model", Scientific American, vol 288(6), p.68-75, 2003

[19] - Jonathan Strickland, how stuff works, http://science.howstuffworks.com/dictionary/astronomy-terms/big-bang-theory7.htm

This phase transition is thought to have happened about 10^{-35} seconds after the creation of the Universe. It filled the Universe with a kind of energy called the vacuum energy, and as a consequence of this vacuum energy density (which plays the role of an effective cosmological constant), gravitation effectively became repulsive for a period of about 10^{-32} seconds. During this period the Universe expanded at an astonishing rate, increasing its size scale by about a factor of 10^{50}. Then, when the phase transition was complete the universe settled down into the big bang evolution that we have discussed prior to this point. This, for example, means that the entire volume of the Universe that we have been able to see so far (out to a distance of about 18 billion light years) expanded from a volume that was only a few centimeters across when inflation began[20]!

5.6.1 Solution of the Problems of the Big Bang by Inflation

If this inflationary epoch really took place, it could cure all the problems of the big bang mentioned above. Briefly;

1- The tremendous expansion means that regions that we see widely separated in the sky now at the horizon were much closer together before inflation and thus could have been in contact by light signals.

2- The tremendous expansion greatly dilutes any initial curvature. Think, for example, of standing on a basketball. It would be obvious that you are standing on a (2-dimensional) curved surface. Now imagine expanding the basketball to the size of the Earth. As you stand on it now, it will appear to be flat (even though it is actually curved if you could see it from large enough distance). The same idea extended to 4-dimensional spacetime accounts for the present flatness (lack of curvature) in the spacetime of the Universe out to the greatest distances that we can see, just as the Earth looks approximately flat out to our horizon. In fact, the inflationary theory predicts unequivocally that the Universe should globally be exactly flat and therefore that the average density of the Universe should be exactly equal to the closure density. It is this prediction that we alluded to earlier when we said that there were theoretical reasons

---

[20] - Matt Strassler, https://profmattstrassler.com/articles-and-posts/relativity-space-astronomy-and-cosmology/history-of-the-universe/inflation/

to believe that the density of the Universe was exactly equal to the critical closure density.

3- The rapid expansion of the Universe tremendously dilutes the concentration of any magnetic monopoles that are produced. Simple calculations indicate that they become so rare in any given volume of space that we would be very unlikely to ever encounter one in an experiment designed to search for them[21].

In addition to (potentially, at least) solving the preceding problems of the big bang, the theory of inflation presents a bonus: detailed considerations indicate that inflation is capable of producing small density fluctuations that can later in the history of the Universe provide the seeds to cause matter to begin to clump together to form the galaxies and other observed structure.

5.6.2 Problems with Inflation

Although inflation has many attractive features, it is not yet a proven theory because many of the details still do not work out right in realistic calculations without making assumptions that are poorly justified. Probably most cosmologists today believe inflation to be correct at least in its outlines, but further investigation will be required to establish whether this is indeed so[22] . In the 1990s, experimental observations showed that the expansion of the universe is accelerating and dark energy is tending to accelerate the expansion of the universe. In March 2014, experimentalists announced that primordial gravitational waves had been discovered. The team behind the BICEP2 Telescope in Antarctica had observed telltale twists and turns in the polarization of the cosmic microwave background radiation (CMB) – the remnants of the earliest light produced in the

---

[21] - Solution of the Big Bang Problems,
http://csep10.phys.utk.edu/OJTA2dev/ojta/c2c/early/inflationary/solution_tl.html
[22] - The Inflationary Universe, University of Tennessee,
http://csep10.phys.utk.edu/astr162/lect/cosmology/inflation.html

universe. Physicists thought the discovery was preliminary confirmation of inflation theory[23] .

5.7 A new definition of singularity

For long time seemed the Friedmann equation is able to explain universe, but in recent years, the cosmological constant was of interest to cosmologists. The Big Bang theory and even the theory of inflation explain after the explosion and say nothing about before the Big Bang. However, these two theories are unable to explain before the Big Bang.

In this book, regarding on review of Newton's second law, author have been attempted to enter to the sub-quantum space by crossing the border of quantum mechanics then to survey of counteracting Newton's second law and the universal gravitation law and finally we can be analyzed and investigated the results. In sub-quantum space, we passed across the black hole and reach the formation of the absolute black hole by specifying the limits of Newton's second law and gravitation law, then the singularity will be explained in the explosion of an absolute black hole. In this review we will be forced to change their attitude towards the singularity and the general conclusion in the singularity state is: *volume will not be zero, density will be limited.*

In this part of the review of singularity, a new definition of singularity provided. But regardless of escape velocity on black hole, we cannot review singularity. If the kinetic energy of an object launched from the Earth were equal in magnitude to the potential energy, then in the absence of friction resistance it could escape from the Earth. Escape velocity is the minimum speed needed for an object to "break free" from the gravitational attraction of a massive body. The escape velocity from Earth is about 11.2 km/s. For a spherically symmetric massive body such as a star or planet, the escape velocity v_{esc} for that body, at a given distance is calculated by the formula:

---

[23] - Michael Slezak, Cosmic inflation is dead, long live cosmic inflation! New Scientist, 2014, https://www.newscientist.com/article/dn26272-cosmic-inflation-is-dead-long-live-cosmic-inflation/

$$v_{esc} = \sqrt{\frac{2GM}{R}} \qquad (5\text{-}10)$$

Where G is the universal gravitational constant, M is the mass of the planet or star, and R is radius of star. If the mass of the star was to such a small size or high density that the magnitude of the escape velocity v_{esc} was greater than the speed of light c, $v_{esc} > c$, then even light could escape the gravitational pull. A black hole has a gravitational field is so strong than not even light can escape its pull[24]. As when a stone is thrown upward to a certain height goes up, the question is, light (a radiation) from the surface of a black hole to how high can go up? The answer is that it depends on the height of the scape velocity at which the black hole. In absolute black hole, intensity of the gravitational field is so large that even light cannot shake from its place at the top. This is just a simple and intuitive definition of an absolute black hole, but we should define an absolute black hole by using the scientific concepts and cosmological equations and analyzing its results. According to equations (1- 37 and 1- 38), energy is formed of sub quantum energy, bearing in mind the equation (1- 34) $E = np_{SQE}c$ and considering the properties of SQEs, we can define the absolute black hole. Here, it is necessary to focus on relations (1- 37 and 1- 38), according to these equations the amount speed V_{SQE} is constant, but the amounts of transmission speed V_{SQET} and non-transmission speed V_{SQES} are not constant, by decreasing the amount transmission speed of SQE is added to the amount non transmission speed and vice versa. Each of these values is maximum when another value is zero that is given by:

$$V_{SQET} \rightarrow V_{SQE} \Leftrightarrow V_{SQES} \rightarrow 0 \qquad (5\text{-}11)$$

$$V_{SQES} \rightarrow V_{SQE} \Leftrightarrow V_{SQET} \rightarrow 0 \qquad (5\text{-}12)$$

---

[24] - Ron Kurtus, "Gravitational Escape Velocity from a Black Hole", The School for Champions, 2011, http://www.school-for-champions.com/science/gravitation_escape_velocity_black_hole.htm#.VxJAOvl97IU

Thus, according to the direction of external force which was affected on a particle/object, the total non-transmission speeds rate is converted to the transmission speeds or to the inverse.

Now we can define an absolute black hole. But before explanations, it is necessary to define two terms of sub quantum divergence and sub quantum converges;

1- Sub quantum Divergence: if a particle/object falls in the gravitational toward a massive body, and the linear speed of its $SQEs$ will be V_{SQET} , we say that the object has sub quantum divergence (figure 5- 1). There is $V_{SQE} = V_{SQET}$ in the sub quantum divergence. So;

$$\text{Sub quantum Divergence;} \quad V_{SQET} = V_{SQE} \Leftrightarrow V_{SQES} = 0 \qquad (5\text{-}13)$$

2- Sub quantum Convergence: if total transmission speeds $SQEs$ of a particle/object go to zero, $V_{SQET} \to 0$, we say that the object has sub quantum convergence (figure5- 1). There is $V_{SQES} \to V_{SQE}$ in the sub quantum convergence. So;

$$\text{Sub quantum Convergence:} V_{SQES} \to V_{SQE} \Leftrightarrow V_{SQET} \to 0 \qquad (5\text{-}14)$$

Definition of an absolute black hole: If a particle/object falls down into the absolute black hole, it will be involved in sub quantum divergence before reaching the surface of the absolute black hole.

Consider the absolute black hole swallowing more matter; its mass and thus its gravitational field intensity will be increase. By increasing the mass, volume is reducing, its constituent $SQEs$ is condensed and its transitional space will be limited. An absolute black hole eats its own gravity effect (gravitons) and nothing even gravitons cannot escape of an absolute black hole.

Definition of Singularity: An absolute black hole with very high density under two followed conditions reaches the singularity state:

1)   constituent $SQEs$ reach sub quantum convergence state i.e. $V_{SQES} \to V_{SQE}$. So the linear speed of everything on the surface of absolute black hole goes to zero, $V_{SQET} \to 0$

2)   Due to the gravitational pressure, the average distance between *SQEs* of an absolute black hole goes to zero. Once the non-transmission speed of *SQEs* reach maximum, $V_{SQES} \rightarrow V_{SQE}$, the average distance between *SQEs* goes to zero due to intensive collision.

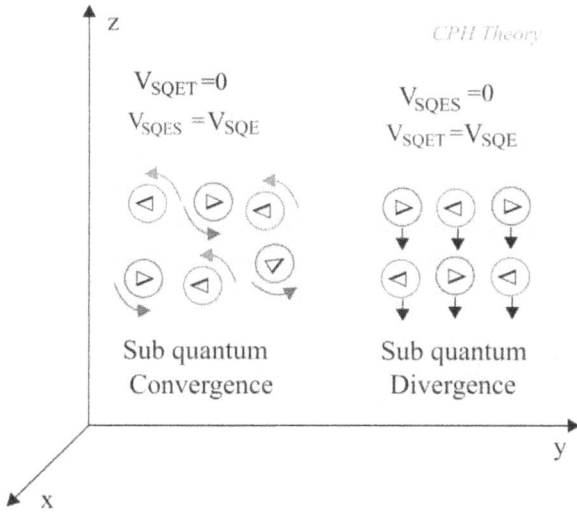

Fig5-1; Sub-quantum Divergence and Convergence

They are scattered around and these chain scattering are spread everywhere inside the absolute black hole and therefore the singularity is occurred. The density is very high in the singularity state, but not infinite. In addition, the volume does not reach to zero, but the average the distance between *SQEs* reach to zero. Given above descriptions can easily explain counteracting Newton's second law and gravity.

5.7.1 Explosion of an absolute black hole

Given the above themes, there are three basic limitations: transmission speed, non-transmission speed and density that they are the reason of creation the observable universe and all physical phenomena existing in it.

Now, by using the equations (5- 13 and 5- 14), the Friedmann's equation (5- 8) that has given as follow and then the Big Bang will be reviewed.

$$\left[\left(\frac{1}{R}\frac{dR}{dt}\right)^2 - \frac{8}{3}\pi G\rho\right]R^2 = -kc^2$$

Right side of the Friedman equation, has given for real space-time and is used for after the Big Bang, because k determined the geometrical properties of space-time and c is the speed of light in a vacuum is constant, but given that the speed of light is not constant in gravitational field (equation 1- 13) and it is zero for surface and inside of an absolute black hole (equations 5- 13 and 5- 14), So if we want to solve the Friedmann's equation for absolute black hole, we must consider the speed of light to zero and the equation becomes as follows:

$$\left[\left(\frac{1}{R}\frac{dR}{dt}\right)^2 - \frac{8}{3}\pi G\rho\right]R^2 = o \qquad (5\text{- }15)$$

Assuming $R \neq o$ (which is a reasonable assumption because the notion that, if the universe collapses, it will not vanish volume and it is not reasonable that universe was created of nothing), then we have:

$$\left(\frac{1}{R}\frac{dR}{dt}\right)^2 - \frac{8}{3}\pi G\rho = o \rightarrow \left(\frac{1}{R}\frac{dR}{dt}\right)^2 = \frac{8}{3}\pi G\rho \qquad (5\text{- }16)$$

We take the square root of the above equation, so we have:

$$\frac{1}{R}\frac{dR}{dt} = \pm\sqrt{\frac{8}{3}\pi G\rho} \qquad (5\text{- }17)$$

$$\frac{dR}{R} = \pm\sqrt{\frac{8}{3}\pi G\rho}\, dt \qquad (5\text{- }18)$$

We take an integral from both sides of above equation:

$$L_n R = \pm\sqrt{\frac{8}{3}\pi G\rho}\, t + C, \qquad C \text{ is integer constant}$$

$$R = e^{\pm\sqrt{\frac{8}{3}\pi G\rho}\,t+C} = e^C e^{\pm\sqrt{\frac{8}{3}\pi G\rho}\,t} \qquad (5\text{-}19)$$

For $t = 0$ the initial radius of the universe is obtained (at the moment of the Big Bang), we have;

$$R_o = e^C$$

For the negative mode, we have:

$$R = R_0 e^{-\sqrt{\frac{8}{3}\pi G\rho}\,t} = \frac{R_0}{e^{\sqrt{\frac{8}{3}\pi G\rho}\,t}} \qquad (5\text{-}20)$$

Equation (5- 20) means that the radius of the universe is shrinking over time and is not acceptable. For positive mode, we have:

$$R = R_0 e^{\sqrt{\frac{8}{3}\pi G\rho}\,t} \qquad (5\text{-}21)$$

Equation (5- 21) is an exponential function that shows in the first moments after the explosion, expansion of the universe was very fast. In addition, because of the big bang, Newton's second law contrasts with the law of gravitational law, in this confrontation, Newton's second law, and the universal gravitational law is neutralized. In the early moments after the Big Bang the speed limit was not the speed of light c, because $SQEs$ collide with each other, everything, even the photons were decomposed and the speed limit could have one of two values SQE speed V_{SQE} or the speed of graviton V_G . So, we can write:

$$\left[\left(\frac{1}{R}\frac{dR}{dt}\right)^2 - \frac{8}{3}\pi G\rho\right]R^2 = -kV_{SQE}^2 \qquad (5\text{-}22)$$

Classical mechanics and relativity (special and general) describe the acceleration is an explanation of outward of phenomena regardless of the properties of sub quantum scales. It should be noted that the interaction between large objects (e.g. collision of two bodies) under the action of the quantum layer (in fact sub quantum layer) done. In sub quantum level, the amount of speed is constant, in any condition and any space, and in any interaction linear momentum changes to nonlinear momentum and vice versa. According to *SQE*, we are able to show there is not a zero volume with infinite density in singularity also before the Big Bang. So, regardless to reconsidering the relativistic Newton's second law, how can we resolve the dark energy problem?

In this book, author tried to show the necessities to investigate about relativistic Newton's second law with different deductions and investigation about some of physical phenomena. Nowadays in some published physical researches, it has been proposed many problems and unsolved questions that will remind with no answer without considering the internal structure of particles and formal relativistic Newton's second law. The correct understanding about the structure of photon enables us to use a new approach on all physical phenomena from subatomic particles to universe. Recently weeks, an article published that shows in the first fraction of a second of the big bang, the Universe was expanded even billions of billions of billions of times faster than today[25]. If we accept the universe is made up which is seeing on the Earth, then explaining the reason of inflation will be simple. In this case, the limit of speed should be the speed of graviton that equation (5- 22) changes to as follows;

$$\left[\left(\frac{1}{R}\frac{dR}{dt}\right)^2 - \frac{8}{3}\pi G\rho\right]R^2 = -kV_G^2 \qquad (5\text{- }23)$$

It is important that we note the speed of graviton V_G was explained before, this speed is so faster than light speed, so that $|V_G| > |V_{SQE}| > |c|$. It is compatible with the inflation theory very well. According to V_G , we are able to describe the cosmic microwave background. The explosion was so severe that a large part of matter of absolute black hole has converted to gravitons. Gravitons were combined and electromagnetic energy was produced. The more amount of electromagnetic energy were converted into

---

[25] - Cosmic inflation: Higgs says goodbye to his 'little brother, EurekAlert, 2017, https://www.eurekalert.org/pub_releases/2017-06/thni-cih060817.php

Original article published in Phys. Rev. D 95, 071101(R) – Published 14 April 2017 https://journals.aps.org/prd/abstract/10.1103/PhysRevD.95.071101

matter and the remaining was formed cosmic microwave background (for the balance of matter and antimatter, see section two).

5.8 Dark matter and dark energy problems

In recent decades, to solve the problems of physics attention to tachyons is on the rise. A tachyon is a hypothetical particle that always moves faster than light. Here are three questions about our universe that modern physics cannot answer:

1 - What is dark energy? Dark energy makes up approximately 68% of the universe and appears to be associated with the vacuum in space[26].

2 - What is dark matter? Studies of other galaxies in the 1950s first indicated that the universe contained more matter than seen by the naked eye[27].

3 - What caused inflation? The blindingly fast expansion of the universe immediately after the Big Bang? For that matter.

The standard model of cosmology indicates that the total mass–energy of the universe contains 4.9% ordinary matter, 26.8% dark matter and 68.3% dark energy[28] (figure 5- 2).

In recent years, several papers have been published to solve dark matter and dark energy problems. The emphasis of these papers is on quantum vacuum and faster than light speed. A new theory proposes that faster-than-light particles known as tachyons could answer a lot of questions about the universe. According to a paper published in European Physical Journal C by Herb Fried from Brown University and Yves Gabellini from INLN-Université de Nice, may be a kind of particle called a tachyon[29].

---

[26] - Voir en français, CERN Document Server, https://home.cern/about/physics/dark-matter

[27] - Nola Taylor Redd, What is Dark Matter?, Space.com, 2017, https://www.space.com/20930-dark-matter.html

[28] - Wikipedia, https://en.wikipedia.org/wiki/Dark_matter

[29] - Robyn Arianrhod, Can faster-than-light particles explain dark matter, dark energy, and the Big Bang?, Cosmos, 2017,

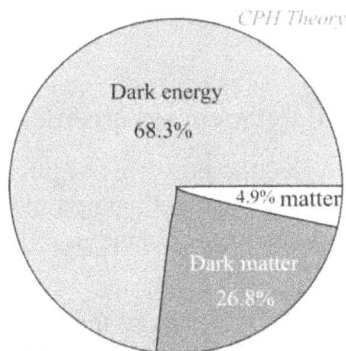

Fig5- 2: Constituent materials in universe

This letter is meant to be a brief survey of several recent publications providing a simple, sequential explanation of dark energy, inflation, and dark matter. These paragraphs lead to an intuitive and qualitative picture of the why and the how of the Big Bang, and thence to a possible understanding of the birth and death of a universe. This review begins by noting a new, QED-based derivation of dark energy that is able to provide the amount of quantum vacuum energy density which astrophysicists believe is responsible for the continuing expansion of our universe[30].

Recent researches show, to solve the problem of dark energy, dark matter and inflation theory, quantum vacuum and faster than light speed should be considered and analyzed, which was done in this book. But regardless to reconsidering the relativistic Newton's second law, how can we resolve the faster than light speed problem? Besides that, the old definition of acceleration prevents the recognition of the nature of acceleration. Due to this reason in 1987, CPH Theory has begun by review the structure of photon, sub-quantum energy and faster than light speed.

---

https://cosmosmagazine.com/physics/can-faster-than-light-tachyons-explain-dark-matter-dark-energy-and-the-big-bang

[30] - H. M. FriedY. Gabellini, The birth and death of a universe, The European Physical Journal C, 2016, https://link.springer.com/article/10.1140/epjc/s10052-016-4577-8

Moreover, one could explain the expansion of the universe better and more real through reviewing relativistic Newton's second law.

6

Physical time and thermodynamics in sub quantum space

Thermodynamics is a branch of physics which deals with the energy and work of a system. Thermodynamics deals only with the large scale response of a system which we can observe and measure in experiments. Small scale gas interactions are described by the kinetic theory of gases.

Our experiences about time come from clocks. Time without clock is never imaginable. Any system with entropy is also a clock.

6.1 Quantum thermodynamics

Quantum thermodynamics is the study of the relations between two independent physical theories: thermodynamics and quantum mechanics. The two independent theories address the physical phenomena of light and matter. But there are some aspects of thermodynamics that are irreconcilable with quantum mechanics. "Quantum mechanically, due to the uncertainty principle, a (more or less localized) free particle always carries a nonzero kinetic energy"[1]. On the other hand, according to the sub quantum energy, the redefined energy seems need to reconsider the thermodynamics laws without using uncertainty principle.

6.1.1 Thermal Energy and the Sub Quantum Energy

In defining the sub quantum energy, we saw that energy and matter are made of sub quantum energy, and the difference between matter and energy depends to their transmission speed. Also, the total amount of transitional and non-transitional speed of sub quantum energy is constant (equation 1-38). So, in CPH theory, from the point of view of speed, energy is same as matter except in transmission speed. In the other word, energy moves with transmission speed of light c, and matter moves with transmission speed of v, that $v < c$. So, speed of heat is c, too, because it is a kind of electromagnetic energy.

The amount transmission speed of matter v is changeable between zero and c, $0 \leq v \leq c$ relative to inertia reference frame, when v become to c (v = c), the matter is converted to energy. It can be shown that the temperature T of the system (such as a gas) is a function of transmission speed of SQEs of system.

Let's assume that a system including k different molecules. In heating the system, the kinetic energy of the molecules increases. In this process, SQEs belong to heat and molecules of system are sharing their momentum and transmission speed of molecules increase (reconsidering the relativistic Newton's second law equation 5- 5). Molecules absorb SQEs of heat, the mass[2] and transmission speed of molecules increase, as thermodynamic interpretation, temperature T of the system increases.

---

[1] - D. Cabreraa, et, at. , ENTROPY, TOPOLOGICAL THEORIES AND EMERGENT QUANTUM MECHANICS, 2017, https://arxiv.org/pdf/1611.07357.pdf

[2] - Tony Rothman, "Was Einstein the First to Invent $E = mc^2$?" 2015,

http://www.scientificamerican.com/article/was-einstein-the-first-to-invent-e-mc2/

6.1.2 Why does material emit energy?

Thermal energy emitted by matter as a result of vibrational and rotational movements of molecules, atoms and electrons. The energy is transported by electromagnetic waves (photons). Radiation requires no medium for its propagation; therefore, it can also take place in vacuum. All matters emit radiation as long as they have a finite (greater than absolute zero) temperature[3]. A system such as gas is made of molecules or atoms, and atoms are not at static state in system. They are moving or oscillating around each other. Also, atoms are made of charge particles, and they absorb or repel each other. So, they are working on each other continuously. In a system, charged particles work on each other and according to section 2, they emit electromagnetic energy. So, every system emits heat energy, and intensity of radiation is depending on its temperature. During the day, the Earth is heated by the Sun and at night, heat escapes from Earth into space. The most important fusion process in nature is the one that powers stars. In the 20th century, it was realized that the energy released from nuclear fusion reactions accounted for the longevity of the Sun and other stars as a source of heat and light. Pressure due to gravity is balanced by pressure of the ionized gas in the star which behaves like an ideal gas. Radiation leaving from the surface determines the luminosity of the star. A star like our Sun has enough hydrogen to fuel the fusion of hydrogen to helium for some 10 billion years during which the Sun shines stably. So system does negative work on itself by converting matter to energy.

6.2 Inherent Power of a System

As each system emits radiation continuously, we can define a work function for every system dependent on the temperature as $W = W(T)$. The higher temperature of system, the more negative work will be done on itself. As a result of this negative work, the system emits heat and its temperature continuously is reduced. The negative work of a system on itself is named the inherent system power and it is shown with P. Inherent power of system on itself is always negative (such as radiation and loses thermal energy), but related to the environment it is positive. This means

---

[3] - Radiation is the transmission of energy in the form of waves or particles through space or through a material medium. Wikipedia; https://en.wikipedia.org/wiki/Radiation

that each system emits heat radiation to environment, even if the system is colder than the environment. It is better to define inherent power of system relative to environment that is positive to help to explain the thermodynamic of the system easier. Relative to the environment view, each system has a positive power $P > 0$ that is defined as follows:

$$P = \frac{dW(T)}{dt} = kE_{SQE} \qquad (6\text{-}1)$$

Where $P > 0$ is inherent power of system and k is a natural number that indicates how many SQE leave the system per time. According to equation (5- 5) system loses energy dE/dt and the kinetic energy of molecules is reduced. Inherent power of system is referred to transfer heat from the system to the environment. Real systems are not isolated and they exchange $SQEs$ with other systems and also the inherent power of a real systems never reaches to zero.

Suppose k_1 is the number of SQE leave system and k_2 enter to system, and $k = k_1 - k_2$ is the result of the heat exchange between system and environment. And if $k > 0$, the inherent power of system is reduced and it cools, such as the pan is removed from the oven. If $k < 0$, the inherent power of system is increasing like a stone under the Sun light. Whenever $k = 0$, two systems are in thermal equilibrium. If P_1 be the primary inherent power of system and P_2 secondary, for an isolated system always: $P_1 > P_2$, and its radiation is decreasing. There is no real isolation system in nature, because the universe itself has radiation which is called the cosmic microwave background radiation. In every direction, there is a very low energy and very uniform radiation that we see filling the Universe. This is called the $2,725°K$ Background Radiation, or the Cosmic Background Radiation.

6.2.1 Thermodynamic Basic-level State of a System and SQE

As already explained, every system has an inherent power that is greater than zero $P > 0$. If a system loses its inherent power, it is at basic-level state of thermodynamics. In the other words, a system would be at basic-level state of thermodynamics, if its inherent power goes to zero $P = 0$ (figure 6- 1).

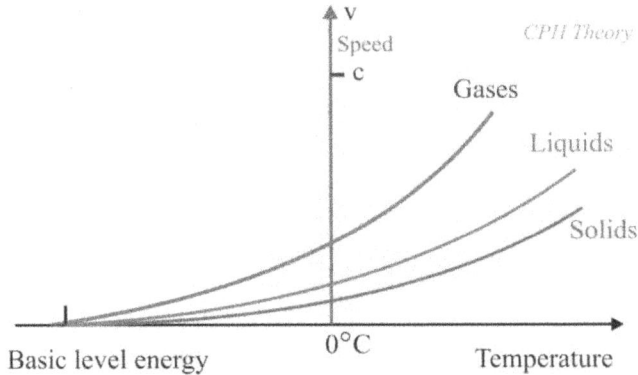

Fig6- 1; speed and temperature of systems

When a system is at basic-level of thermodynamics, its charged particles are not able to work on each other, so the system does not emit heat energy. When a system is at basic-level of thermodynamics, then its temperature is absolute zero. Suppose a system is at basic-level of thermodynamics, it contains n of *SQE*s that are moving with velocity $v_1 = 0$ in system. We give heat to it, in fact k of *SQE*s with speed c enter the system, and particles of system absorb them. In a real environment the inherent power of a system cannot be zero, even in space, because there is cosmic background radiation in space. A real system never reaches to basic-level of thermodynamics, because no system in the universe can be absolutely isolated.

6.2.2 Thermodynamics laws and sub quantum energy

First Law of Sub Quantum Thermodynamics

A system works on the environment with inherent power $P > 0$. To stabilize or increase the internal energy of a system, we must give heat energy to system with power $P' \geq P$.

The Second Law of Sub Quantum Thermodynamics

There is no actual physical process by which we can make the inherent power of a system P one-way. Consider that in an actual physical process the inherent power is not constant. Because heat energy incomes and outgoes of an actual systems.

Third Law of Sub Quantum Thermodynamics

An actual physical system never approaches the basic-level of thermodynamics. There is no physical process to take a system to the thermodynamic basic-level state.

Sub Quantum Entropy

Entropy (ΔS) of a system is equal to its inherent power (P), $\Delta S = P$, so entropy of a system approaches zero only at basic-level thermodynamics.

6.3 Physical time

Time is one of the most complex concepts that usually the human mind has been preoccupied about it. Scientists and philosophers have struggled to identify and explain the nature of time. However, still there is no physical definition of time, and it is still just a scientific undefined quantity. "It would be nice if we could find a good definition of time." Richard Feynman said[4]. With naivety about time, time is passing from past to the future which from all eternity would last forever. This is exactly the simplified form of the absolute time of Newtonian physics. Although, the nature of time was intended, but do we really have anything for time except the clock and its ticking. Richard Feynman once quipped, "Time is what happens when nothing else does[5]." If nothing happens, so there is nothing cause of happen,

---

[4] - The Feynman Lectures on Physics Vol. I Ch. 5: Time and Distance
http://www.feynmanlectures.caltech.edu/I_05.html

[5] - Julian Barbour, "The End of Time: The Next Revolution in Physics",
http://www.amazon.com/The-End-Time-Revolution-Physics/dp/0195145925

when there is nothing, what is time that could happens? So it seems what Feynman has said is a philosophy explanation (not defined) of time and it is not even physical explanation. However, Julian Barbour disagrees with Feynman and says: "if nothing happened, if nothing changed, then time would stop. For time is nothing but change. It is change that we perceive occurring all around us, not time. Put simply, time does not exist."[6] Some other authors and researchers say that time is an illusion[7]. "The meaning of time has become terribly problematic in contemporary physics, the situation is so uncomfortable that by far the best thing to do is declare oneself an agnostic." says Simon Saunders[8].

Efforts to understand time below the Planck scale have led to an exceedingly strange juncture in physics. The problem, in brief, is that time may not exist at the most fundamental level of physical reality. If so, then what is time? In addition, why is it so obviously and tyrannically omnipresent in our own experience?[9]

In order to have better understanding of the physical nature of time, reviewing physical properties of particles can be helpful and how they interact. The question is: what is the physical nature of time? Moreover, which particles do not experience passing time? Are there such particles? If so, what are their features? In physics, we see only clocks, our efforts is to understand more than ticking clocks. Is there time independent of clocks? Is time describable without of a clock? Usually, clock is introduced as a tool that shows passing time, behind this simple definition of clock, there is something else. This ambiguity permeates to thought that not all clocks are synchronized with each other, and surprisingly they have very different lifetime. Probably human is looking for an eternal clock, a clock that works independently of all physical events and will work forever.

---

[6] - Last reference. Julian Barbour,

[7] - Craig Callender, "Is Time an Illusion?", Scientific American, 2010,

[Online] available; http://www.scientificamerican.com/article/is-time-an-illusion

[8] - Folger, Tim, "Time May Not Exist" 2007, Discover,
http://discovermagazine.com/2007/jun/in-no-time

[9] - Last reference, Folger,

6.3.1 A commonplace definition of clock

A clock is an instrument to indicate, keep, and co-ordinate time. (In fact, an instrument reveals or observes iteration of a specific event). The Earth is a clock, because it continuously shows the iteration of sunset and sunrise. Heart of human being is like a clock, because its beats is an iterative event. However, certain clocks are more regular than some other clocks. In comparison, the Earth is a clock more regular than heart of human being. Because by running or sickness, the rate of heart beating changes. Therefore, we cannot take into account Earth and heart as two synchronous clocks, because sometimes the distance between two sunrises happens x times and sometimes with y times of heart beating. Of course, it might be said that this is not a failure or deficiency of heart beating, it is from the Earth that irregular with respect to heart, but we have other clocks that we can compare with it, like sand clock.

We can make a sand clock and show that it is synchronous with Earth clock, but it is not synchronous with heart clock and again we make a mechanical clock and compare it with Earth and heart and show that the problem is from heart clock. Therefore, some of the clocks are more regular than other clocks and most well-known regular clock is atomic clock[10].

Is there any special physical relation between clock and iterative event that indicates passing time? Because in physics, we cannot follow nonphysical communications, so this communication must be specified by investigating physical phenomena, if there exists.

6.3.2 Physical nature of time

In equations and computation of classic physics, time used an absolute quantity. The subject that time is really absolute or not, did not discuss anymore and it was not considerable to draw attention of researchers and scientists. Because, there was not any physical experience that be inconsistent with being absolute of time. However, after "Michelson-

---

[10] - A Brief History of Atomic Clocks at NIST,
http://tf.nist.gov/cesium/atomichistory.htm

Morley Experiment"[11] and formulation of "Lorentz Transformation"[12] and especially with appearance of relativity, synchronization of two same clocks was challenged in different speeds and more than even the physical nature of time was questioned of. However, in relativity, it has been discussed about dilation of time. Michelson-Morley experiment was incompatible with the Galilean transformation but totally compatible with the classical principle of relativity under the Lorentz transformation. The incompatibility of the laws of electromagnetism, the classical principle of relativity, and the Galilean space-time transformation[13] led Einstein to a critical reevaluation of the concepts of space, time, and simultaneity[14]. Nevertheless, in relativity just motion of clock is noteworthy and under investigation. For example, pay attention to definition of proper time in general relativity:

A proper time is a time that an observer that moves in space-time, measures by means of his/her clock. Proper time has a high importance in general relativity, because observer can measure the effects of time dilation in different locations and various paths that goes through in space-time by using it[15]. Essentially, in relativity, it is not discussed about physical nature of time. Discussion about nature of time was proposed after philosophy in quantum mechanics.

Quantum mechanics investigates about time from both experimental and conceptual dimensions. Although paradox of twins was proposed in

---

[11] - Leonardo Motta, "The Michelson-Morley Experiment",

http://scienceworld.wolfram.com/physics/Michelson-MorleyExperiment.html

[12] - Lorentz Transformation Equations,

http://ffden-2.phys.uaf.edu/212_fall2003.web.dir/Eddie_Trochim/Lorentztransform.htm

[13] - Space-Time Structure, Newton's Laws and Galilean Transformation

http://www.newagepublishers.com/samplechapter/000485.pdf

[14] - Marshall L. Burns, "MODERN PHYSICS FOR SCIENCE AND ENGINEERING" Electronic Textbook License Agreement, 2012 , pp. 11 and 36

http://web.pdx.edu/~pmoeck/lectures/Modern%20Physics%20for%20Science%20and%20Engineering%20(eval).pdf

[15] - Time measured by a clock that has the same motion as the observer. Any clock in motion relative to the observer, or in a different gravitational field, will not, according to the theory of relativity, measure proper time. The free dictionary; http://www.thefreedictionary.com/proper+time

relativity, but quantum mechanics claims, photon does not experience passing time[16]. Photon travels in space milliards of light years and when it enters to light systems on the Earth, behaves like photons in which have been created on the Earth and in the lab a moment ago. Is photon stationary (static) on the axis of time? In other words, is photon moving in the space that does not experience passing time in the words of quantum mechanics, but motionless in the time? It means that here this question that, is not there a space-time to the relativistic concept for the photon.

Generally, the mass (or energy) with space, movement and time are intertwined and are inseparable which other that we can observe, detect or visualize a physical being. Image or visualize a physical entity in our mind is something and its physical reality is something else. If these two (phenomenon and physical reality) were unit, we did not need to research and think about the reality of a physical entity and everything was clear for us, but it is not. Our knowledge of the physical beings is the result of centuries and continuous efforts that we have reached to modern physics. There is three attitudes about time in the modern physics. One is based on relativity, other is a result of quantum mechanics progress and the third attitude of thermodynamics. In all three views, space, mass (or energy) and movement are the intrinsic properties of a physical entity, but there is considerable differences view about time.

In special relativity, moving clock works slower than static clock. What does it mean moving clock in special relativity? It means that, the clock A that is in inertial system and moves with the speed v, works slower than the time that was in the same static system. This proposition has been proved in experience with life-span of peon and muon[17]. We can make a conclusion from this discussion that if transmission speed of clock increases, it works slower. This experience can be compared by using photons, photon moves with transmission speed equal to speed of light and according to emphasis of quantum mechanics; it does not experience passing time.

---

[16] - William and Deborah Whillyard, Physics - Standard Model
http://www.whillyard.com/science-pages/electromagnetic-force.html
[17] - Cosmic ray muons and relativistic time dilation, Preston College Cosmic Ray Group,
https://teachers.web.cern.ch/teachers/archiv/hst2000/teaching/expt/muoncalc/lifecalc.htm

According to general relativity, the clock lies in a stronger gravitational field, is slower than a clock that works in a weaker gravitational field. The gravitational time dilation has been verified by experience[18]. Till here, we have two events that can lie the time in relation with physical conditions of a system (or object). However, we have another event that relies on experience more than it can be relied on theory and essentially its rules has experimental base and it is thermodynamics. In thermodynamics, entropy is known as arrow of time[19]. If we consider the visible universe as an isolated system, irregularity is getting increased at the moment of big bang till now. Therefore, entropy of the universe can be used as a time arrow.

So far, we have four experimental samples about physical time, speed is associated with time in special relativity, In general relativity, time is affected by gravity, in quantum mechanics is claimed that there is not time in quantum scale[20] and it is measurable in thermodynamics of entropy with the arrow of time. From the standpoint of feeling, time has an arrow from past to future. This attitude is consistent with classical mechanics and thermodynamics, but is not compatible with relativity and quantum mechanics. Now let's see how these four physical realities can be described using the sub quantum energy SQE. However, we have another amazing quantum mechanics reality that is quantum entanglement. Entanglement occurs when a pair of particles, such as photons (or electrons), interact physically. A laser beam fired through a certain type of crystal can cause individual photons to be split into pairs of entangled photons. The photons can be separated by a large distance, hundreds of miles or even more. When observed, photon A takes on an up-spin state. Entangled photon B, though now far away, takes up a state relative to that of photon A (in this case, a down-spin state)[21] . That was what made Einstein as "spooky actions at a

---

[18] - Edwin Cartlidge, "Gravity's effect on time confirmed", Physics World, 2010, http://physicsworld.com/cws/article/news/2010/feb/17/gravitys-effect-on-time-confirmed

[19] - ANDREW GRANT, "The arrow of time", Science News, 2015, https://www.sciencenews.org/article/arrow-time

Joel L. Lebowitz, "Time's arrow and Boltzmann's entropy", 2008, http://www.scholarpedia.org/article/Time's_arrow_and_Boltzmann's_entropy

[20] - Quantum Theory Proves That Time Does Not Exist, 2013, http://wokenmind.com/quantum-theory-proves-that-time-does-not-exist/17/03/2013

[21] - Karl Tate, "How Quantum Entanglement Works (Infographic)", 2013,

distance" refers, since changes were occurring at a point immediately at another point[22].

How this event can be explained? This event unlike thermodynamic scale of time, happens in the viewpoint of CPH theory out of time dimension and it is something beyond of time dilation. The question is: Where is the drawback of different interpretations of physical events and inconsistency in theories with each other? Is nature mysterious or unknowable? The reality is another thing. The reality is that nature rules and physical behavior of existents has been in the form that is now too. Just our knowledge and our recognition has grown and our insight has permeated from surface to depth. For better understanding and explanation of these events and different opinions, we need to permeate more in the depth of nature and take some distance of outward (appearance) of events. It means that we need to pass from the boundary of the apparent world (speed of light) and simultaneous with passing from the boundary of the apparent world, we must leave quantum scales and analyze the processes in sub quantum scales in which relativity and quantum mechanics are not able to explain them. Quantum mechanics and relativity work in quantum scales and high speeds near to speed of light, but they are unable to explain beyond that. The problems of modern physics is due to this reason that these theories have been stopped in the boundary between speed of light and faster than light and also in quantum scales. However, physical realities like vacuum energy and virtual photons indicated that speed of light and observable particles is not the end of physical spaces.

6.4 Fields entanglement

Let's look at gravitational blueshift; when a photon is falling in a gravitational field, its energy and frequency increase. Regardless of the

---

http://www.livescience.com/28550-how-quantum-entanglement-works-infographic.html

[22] - Charles Q. Choi, "Quantum Entanglement Benefits Exist after Links Are Broken", Scientific American, 2009

http://www.scientificamerican.com/article/quantum-entanglement

Tia Ghose, "Spooky Action Is Real: Bizarre Quantum Entanglement Confirmed in New Tests", 2015 http://www.livescience.com/52811-spooky-action-is-real.html

atmosphere and according to structure of photon, three fields are entangled in gravitational blueshift, gravity field that is made up of gravitons, virtual field that contains virtual photons and real electromagnetic field of real photon (see section one). Each field has its own space that is describable as follows:

1- Real space-time; everything moves with speed v $\leq c$ in real space-time. Light speed is the highest speed in the real space-time.

2- Virtual space-time; it is called sub quantum energy (SQE) too. Every particle such as virtual particle is explainable in the virtual space-time. Every virtual particle moves with speed V_{SQE}, so that $V_{SQE} > c$. Virtual particles and their physical properties is explainable in virtual space-time without using uncertainty principle.

3- Non-obvious space (NOS); everything like graviton is not directly (also indirectly) detectable in non-obvious space. But, their existence and properties can be found of their effects. Productions of this space are sub quantum energies and virtual photons.

In fact, graviton is converted to sub quantum energy and virtual photon is formed of sub quantum energies. A real photon is obtained with combination of positive and negative virtual photons in which its mechanism was already explained (equations 1- 30 to 1- 35).

6.4.1 A look at the interplay of three spaces

Our views, ideas and physical experiences are limited to observable universe or real space-time, because our tools and we belong to real space-time and obey of its rules. That is why we cannot see or detect virtual existents. However, we can see their effects.

World lines of a real photon and a virtual photon in Minkowski[23] space-time is the edge of real space-time where real photon is moving in vacuum

[23] - John D. Norton, "Spacetime",
ttp://www.pitt.edu/~jdnorton/teaching/HPS_0410/chapters/spacetime/

Nilton Penha and Bernhard Rothenstein, "Special Relativity properties from Minkowski diagrams", http://arxiv.org/ftp/physics/papers/0703/0703002.pdf

with speed c, in this edge, vacuum energy is produced and appears. Thus, in the edge of Minkowski space-time, electromagnetics and gravity are unified with each other. We consider a small cutting in the structure of the photon and its surroundings in a vacuum and investigate its mechanism in the boundary of Minkowski space-time. In this small incision real space-time, virtual space-time and non-obvious space are involved with each other (figure 6- 2).

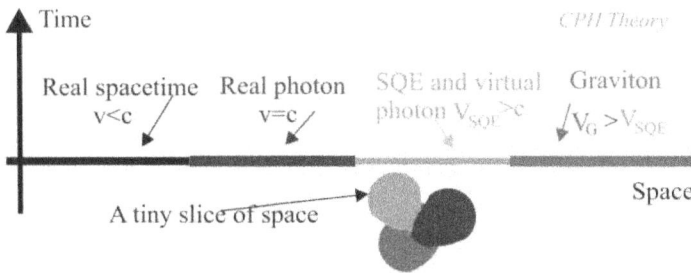

Fig6- 2; photon, virtual photon and graviton are entangled in a tiny slice of space

It is notable that these spaces are indivisible of each other, they are closely intertwined. Any small slice of the available space is composed of above three spaces. Interaction between these spaces causes creation and annihilation the detectable particles.

6.4.2 Minkowski formula and SQE

Here, we concentrate on speed and momentum of real and virtual photons. Therefore, we use light-like interval that given by[24]; $c^2t^2 = r^2$ or $S^2 = 0$.

---

[24] - Susskind Lectures, Space-time separation, http://www.lecture-notes.co.uk/susskind/special-relativity/lecture-4/space-time-separation/

World lines of particles and NR-particles relative to an inertial observer in (x, y, z, t) frame (according to argument and based on physical experiments) in Minkowski space-time can be written as follows[25]:

$$Real\ spacetime;\ x^2 + y^2 + z^2 = c^2 t^2 \qquad (6\text{-}2)$$

$$Virtual\ spacetime;\ x^2 + y^2 + z^2 = V_{SQE}^2 t^2 \qquad (6\text{-}3)$$

$$Non - obvious\ space:\ x^2 + y^2 + z^2 = V_G^2 t^2 \qquad (6\text{-}4)$$

Equation (6- 2) shows world line of photon that lies in the border of real space-time; world line of other particles such as electron that moves with speed $v < c$, is given by;

$$Particles\ worldline;\ x^2 + y^2 + z^2 = v^2 t^2, v < c \quad (6\text{-}5)$$

World line of other physical extents including virtual photon and graviton is out of real space-time. In equation (6- 3), when $V_{SQE} = c$, virtual particles appear in real space-time in which it is indirectly detectable in the structure of photon. When $V_{SQE} < c$ it is a part of quantum particles such as electron (equation 2- 30). The boundary between real space-time and virtual space-time is speed of light c. In gravitational blueshift and zero point energy; virtual photons leave virtual space-time and enter into the real space-time. Also in gravitational blueshift, at first gravitons enter from non-obvious space into virtual space-time, then pass from virtual space-time and enter into real space-time (figure 6- 3).

According to relation $|V_G| > |V_{SQE}| > |c|$, every visible (detectable) physical existent is decayed. Here, meaning of decay is not necessarily spontaneous decay but it may be converted to other particles. Also each virtual particle is decayed, but graviton does not decay, thus it does not

[25] - J. M. CHAPPELL, et. at. (2011). Revisiting special relativity: A natural algebraic alternative to Minkowski spacetime. arXiv:1106.3748v1 [physics.class-ph] .

Bros, J. (2005). The Geometry of Relativistic Spacetime: from Euclid's Geometry to Minkowski's Spacetime. Service de Physique Th´eorique, C.E. Saclay (pp. 1-45). Gif-sur-Yvette, France: Online available, http://www.bourbaphy.fr/bros.pdf.

experience "passing time". If graviton does not experience passing time, this question arises what does means parameter of time t in equation (6- 4)?

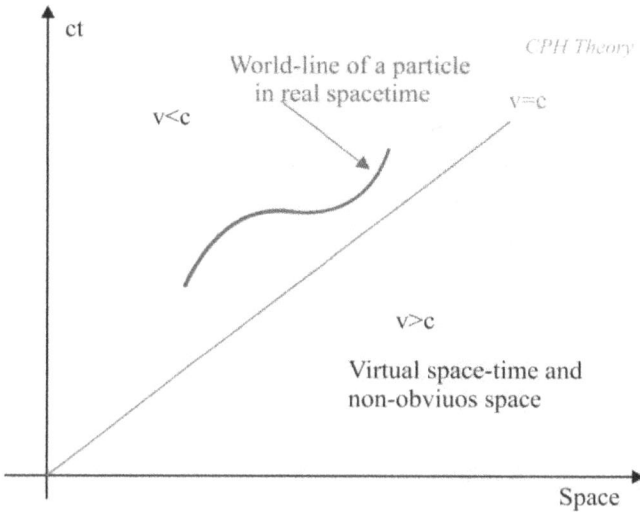

Fig6- 3: world lines of particles in the Minkowski diagram

This equation is an assumption, for an inertial observer in real space-time. The above equation is not the only possible option and unique, the imaginary of Murkowski's formula has discussed[26]. If a graviton writes its own world line equation, it may be same as ; $x^2 + y^2 + z^2 = 0$, for a graviton $t = o$, because a graviton never is created or decayed. By solving this equation in imaginary space we have;

$$x^2 = (-1)(y^2 + z^2) = i^2(y^2 + z^2)$$

$$x = \pm i\sqrt{y^2 + z^2}$$

[26] - Chappell, J. M., Iqbal, A., Iannella, N., & Abbott, D. (2012). Revisiting special relativity: A natural algebraic alternative to Minkowski spacetime. *PloS one, 7*(12), e51756

$$x^2 = (-1)(y^2 + z^2) = i^2(y^2 + z^2) \qquad (6\text{-}6)$$

$$x = \pm i\sqrt{y^2 + z^2}$$

The life of graviton is independent of time. It exists in an imaginary space and moves, so that it is not observable and detectable for human that lives in a real space-time. Moreover, graviton carries information and moves so much faster than light speed c. According to properties of color-charges and magnetic-colors G, G^-, G^+, G^m, graviton carries pure information in which it moves with infinite speed from the viewpoint of an inertial observer. Such an observer cannot measure the motion of graviton and transmission of its information with its clock that obeys the rules of real space-time.   It is considerable and comparable with the above description that some physicists claim; everything happens at same time[27]. In addition, author of this book has expressed same point of view and that was published in Persian language in 1992[28].

6.5 Quantum entanglement

Regard to quantum entanglement this question arise; is quantum entanglement connected by a physical being between particles? Or non-physical? If it is a physical being, then what is it? As it has mentioned in section one, not only massive particles such as electrons, even photons carry their own gravitational fields that are inherent in their mass-energies properties. But the universal connection between everything, also particles is gravity (in fact gravitons) only.

---

[27] - SEAN MARTIN, Time is NOT real: Physicists show EVERYTHING happens at the same time, 2016

http://www.express.co.uk/news/science/738387/Time-NOT-real-EVERYTHING-happens-same-time-einstein

[28] H. Javadi, "Unified Force, Mass and Energy", Scientific-Technical Journal of Islamic Azad University, South Branch, vol. 1, pp. 15- 17, 1991, For more read; New reasonable evidence of CPH Theory's predictions

https://www.researchgate.net/publication/311962908_New_reasonable_evidence_of_CPH_Theory%27s_predictions

Relativity and quantum mechanics are typically tested are widely separated, their foundational principles are rarely jointly studied[29]. Due to this reason, there is dubiety that entanglement is the essence of quantum weirdness or it is the essence of space-time geometry[30]. In fact graviton (also, color charges and color magnet) is carrying pure information that for a real observer it moves with infinite speed.

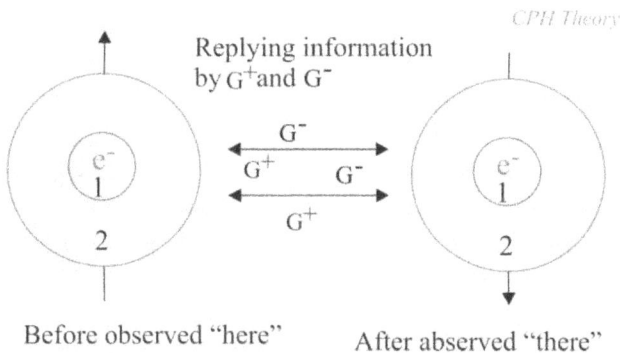

Fig6- 4; Quantum entanglement

All particles are steady exchanging gravitons (and color - charges) that move faster than light speed. When a color - charge reaches to particle A from particle B, particle A reacts to particle B, same happens for particle B, so they are entangled.

As well as, the entropy of a system is spreading of information. A star radiates due to its inherent power in which leads us to notice its existence and physical features. The information related to a star can be revealed through the photons, which it emits. Therefore, we can reveal and understand for a photon both its existence and properties when either we directly observe it by a real photon or we can reveal it through a virtual photon which is emitted by an electron. These revelations are related to the

---

[29] - Igor Pikovski, et, at. Time dilation in quantum systems and decoherence, 2017,
http://iopscience.iop.org/article/10.1088/1367-2630/aa5d92

[30] - Ron Cowen, The quantum source of space-time, 2015,
http://www.nature.com/news/the-quantum-source-of-space-time-1.18797

obvious universe, but the information related to the existence and properties of fundamental particles are also propagated by gravitons with speed faster than light speed. As we know, the following elements are not observable for us, because:

$$V_G, V_{G+}, V_{G-}, G_m^+, G_m^- > c \quad (6\text{-}7)$$

Therefore, quantum entanglement is explainable in a non-obvious space and is not limited to two particles, it could be included to a large number of particles[31]. An important conclusion from this discussion will be as follows:

In an obvious universe, physical time does not exist, independent of matter (energy). Whenever "time" is involved, one clock is associated, because human is also a physical existence and consequently he/she is a clock, too. On the other hand, a physical existence (able to being obvious) is made in its own space, and the moment that it is created, its time starts. Therefore, man is a clock, too and when we talk about the nature of time, apart from comparing the rhythm of the movement of the clocks, nothing else is explainable in physics.

According to the above mentions and equation (6- 2 to 6- 5) with a definition, we present a general rule of objects/particles that experience or do not experience passing time and by using it we will be become closer to recognize the physical nature of time. In sub quantum thermodynamics, we saw that any physical system with inherent power $P > 0$ works on the physical environment. Now we define the physical time by using sub quantum inherent power and investigate it in relation with time in relativity, thermodynamics and quantum mechanics.

6.6 The sub quantum energy and time

Any physical being does not keep its content energy in interaction with other physical entity, experience passing time. By this definition any

---

[31] - Xi-Lin Wang, Experimental ten-photon entanglement,
https://arxiv.org/ftp/arxiv/papers/1605/1605.08547.pdf

physical entity that is keeping energy content, does not experience passing time, in other words, its existence is independent of time.

6.6.1 Quantum mechanics and time

In quantum mechanics, all particles have variable content of energy (except graviton), even photons can gain energy in interaction with other particles, like as inverse Compton scattering[32] , and the gravitational blueshift, or lose energy as Compton scattering and gravitational redshift. Therefore, energy of photon changes. Most importantly, in pair production, a photon is converted to "electron-positron" pair, in this process, photon is decayed and "electron-positron" pair are created. These are all reasons that the photon will experience passing time, but in a long journey in space, it moves with limit speed c, as long as it does not interact with other particles or fields, it preserves its content of energy and in this travelling , does not experience passing time. However, when photon interacts with other particles or fields, experiences passing time. Even photon while in escaping from a black hole, loses all its energy and its life ends. Thus photon that moves in real space-time with speed of light, experiences passing time, besides other quantum particles experience passing time. It means time exists in quantum mechanics and all particles have a finite lifetime.

6.6.2 Relativity and time

In special relativity, moving clock works slower than stationary clock. Time dilation in special relativity must be investigated along with contraction of length. Because they are inseparable from each other. Contraction of a physical object means compactness of atoms and sub atomic particles. Whatever atoms are compacted more to each other, the inherent power of system decreases for any reason that is considered. Consider a radioactive element instead of a clock, radioactive elements in high speeds radiate less than low speed. A bit more precisely: "An observer on station A measures time using his on-board clock. Station B, passing A

---

[32] - Inverse Compton Scattering,
http://eud.gsfc.nasa.gov/Volker.Beckmann/school/download/Longair_Radiation3.pdf

at high speed, has an exact copy of A's clock on board. Yet, from the point of view of A, the clock in station B runs more slowly than his own. A down-to-earth version of this effect can be tested with the help of elementary particles such as those particles accelerated inside the "proton synchrotron" of the European research Centre CERN"[33].

In addition, here simultaneously two relativistic effects, expansion of time and contraction of length must be considered, reduction of volume and dilation of time have direct relation with each other. Because due to reduction of volume, inherent power of radioactive elements decreases. In addition, time dilation in general relativity happens by reduction of volume, due to gravitational pressure. The clock that is on the Earth is under gravitational pressure more than a clock that lies at the top of a mountain.

Time dilation in proximity of black hole is more than surroundings of a luminary like a planet and even the time stops in the horizon of an event of a black hole (from the viewpoint of an external observer). Volume decreases due to gravitational pressure and inherent power of objects and particles reduces. From the viewpoint of an external observer, the time stops completely in a black hole. Therefore, there exists a direct relation between inherent power of system and gravitational pressure, in which it causes dilation of time in general relativity.

It must be noticed that the performance of gravitational pressure is limited. Therefore, however there exists time dilation in general and special relativity, however, it does not tend to zero for any object and particle, it means that all physical existents experience passing time in general and special relativity.

6.6.3 Thermodynamics and time

From viewpoint of sub quantum thermodynamics, any system with inherent power P works on the physical environment, so its content of energy is not constant. Therefore, from thermodynamic viewpoint all systems experience passing time. Then it is acceptable that by thermodynamic

---

[33] - Einstein online, The relativity of space and time,
http://www.einstein-online.info/elementary/specialRT/relativity_space_time

insight to the time, thermodynamic time is oriented and from the past to the future.

In thermodynamics that its study domain is large systems (relative to quantum particles), any physical existent (system) has its special location on the time axis from the past to the future. On the other hand, in thermodynamics, heat exchange of systems are studied in observable universe. The observable universe is a real space-time where it's speed is always smaller or equal the speed of light, in such speed, always we have: $v \leq c$. The time is a real quantity and it should be used in computations in thermodynamics.

Differences between response of quantum mechanics and thermodynamics to the nature of time is referred to the difference of their expertise area. Expertise area of quantum mechanics is to recognize the nature of interactions and properties of radiations, while expertise area of thermodynamics is radiation systems and relation between them.

In quantum mechanics, some particles (such as photon) do not experience passing time, but in thermodynamics, any thermodynamic system (from a capsule of gas until observable universe), either have oriented time axis for themselves from the past to the future and the time never stops.

6.7 The sub quantum definition of clock

According to CPH theory, everything (except graviton) has an inherent power $P > o$ radiation continuously energy (electromagnetic radiation) and this process is repeated. That means in real space-time everything is a clock. If we consider the objects as a clock, the ticking of clock is radiation objects. Some objects are so irregular that we do not use them as a clock, and someone are regular clocks than others. In quantum mechanics, a photon is a unit of radiation (in terms of quantum mechanics) and does not experience passing time. When this proposition is acceptable that the energy of photon does not change and it was an unstructured particle, while it is not and energy of the photon also changes. But this attitude of quantum mechanics is a good guide for understanding the physical time. To consider the following hypothetical experience. Let's assume that a person in the laboratory is producing pair electrons - positrons by shining high-energy photons on a leaden sheet. Someone else other side of laboratory is combining the particles - antiparticles and generate new photons. Third party is out of laboratory is seeing the photons have been created by combination particles - antiparticles and he is unaware of what is happening in the lab. Until the twentieth century, human was like the third person in the above assumption. He did not know that mass is convertible to energy and vice versa, but we know.

Assume that the observable universe would collapse due to gravity. Again, a new universe appears by another big bang. We suppose a smart existent like human lives in the next universe, the question is: How he/she will know that we have lived before him/her? All the materials in the observable universe converted to energy and energy converted to matter again. How do we know there was a universe before the present universe or it has not been existed at all? We do not know the answer of this question. However, we know that any physical existent in this universe does not destroy and just it converts to another thing, converting energy to mass and vice versa, in fact, field converts to energy, energy converts to matter-antimatter and vice versa.

The fact is that the Earth, Solar system and the universe existed before us and after us will exist too. Human as a clock compares himself with older clocks, and from this comparison concludes that there exist something that is called time, and the time is independent of physical existents. While any attempt to explain or define time, independent of physical existents has been inconclusive. A physical existent with its own space and time forms its special shape and after a number of ticking, decays or converts to other physical existence.

At the lowest level of physical universe, in a small slice of space (in quantum vacuum), the three spaces, real space-time, virtual space-time and non-obvious space are intertwined and "quantum energy" is appeared. Just for non-obvious space, time does not exist, and beyond the non-obvious space, everything experiences passing time. Is there another universes beyond the observable universe[34]? Alternatively, is the boundary of

_____

[34] - Josh Clark, "Do parallel universes really exist?" HowStuffWorks, http://science.howstuffworks.com/science-vs-myth/everyday-myths/parallel-universe.htm

observable universe, the end of being? In this case, there are a lot of debates and discussions, but this question is such a question proposed many years ago long before the invention of the telescope, is there something beyond the Solar system? At the beginning of the twentieth century, we did not know what there is beyond the Milky Way. Can we nowadays see beyond the observable universe? However, there is a thinkable question, all the stars are radiating electromagnetic energy or photons that emit to all around, so what happens for these photons that reach to boundary of visible universe? Is it absorbed before they reach to the border? If so, what does it attract them? It seems that there is nothing, which absorbs all photons in the observable universe. Is there a dam or barrier in boundary of universe to prevent their leaving? If so, what is it? In addition, there is no reason that no radiation reaches to our universe from the outside. We cannot show or prove except the visible universe, there is another universe or not! Maybe we should wait to develop technology in the future, so that we can analyze outside of the visible universe.

6.8 Coefficient of entropy-lifespan

According to relations (6- 2 to 6- 6) and thermodynamic inherent power $P > o$, it can be defined the entropy-lifetime coefficient for all physical existents from sub quantum particles to the largest physical systems as follows:

$$C(\Delta S, t) = \frac{m}{M} \qquad\qquad (6\text{-}8)$$

Where $C(\Delta S, t)$ is the entropy-lifetime coefficient and dimensionless, ΔS is entropy, t is the quantity of time, M is the mass of particle or system and or any physical existennce and m is the mass of radiation due to thermodynamic inherent power $P > 0$. Moreover M must be stable (invariant). If $m = 0$ then the above two conditions holds for graviton, because M is a constant and m is equal to zero and it does not experience passing time. But as far as experience shows us, other particles and physical systems is violated one of following conditions for graviton, because m is not zero, or M is not stable. Moreover, since graviton is a base and formed of structure of energy and matter and does not experience passing time.

Therefore, universe (obvious and non-obvious universe) does not have any beginning and any ending, it means that from the altitude of time, the universe has no pre-existence and no eternity and they are coinciding (matching) on each other. In more clear expression, we can imagine for universe "no pre-existence and no eternity ".

www.ingramcontent.com/pod-product-compliance
Lightning Source LLC
Chambersburg PA
CBHW031121210326
41519CB00047B/4230